BASIC HEALTH PUBLICATIONS USER'S GUIDE

TO CAROTENOIDS & FLAVONOIDS

Learn How to Harness
the Health Benefits
of Natural Plant
Antioxidants.

JACK CHALLEM AND
MARIE MONEYSMITH
JACK CHALLEM Series Editor

The information contained in this book is based upon the research and personal and professional experiences of the authors. It is not intended as a substitute for consulting with your physician or other healthcare provider. Any attempt to diagnose and treat an illness should be done under the direction of a healthcare professional.

The publisher does not advocate the use of any particular healthcare protocol but believes the information in this book should be available to the public. The publisher and authors are not responsible for any adverse effects or consequences resulting from the use of the suggestions, preparations, or procedures discussed in this book. Should the reader have any questions concerning the appropriateness of any procedures or preparations mentioned, the authors and the publisher strongly suggest consulting a professional healthcare advisor.

Series Editor: Jack Challem
Editor: Laura Jorstad
Typesetter: Gary A. Rosenberg
Series Cover Designer: Mike Stromberg

Basic Health Publications User's Guides are published by Basic Health Publications, Inc.

ISBN: 978-1-59120-140-3 (Pbk.)
ISBN: 978-1-68162-845-5 (Hardcover)

CONTENTS

INTRODUCTION

In recent years, the phrase *eat plenty of fruits and vegetables* has become a mainstay in the world of nutrition and health. By now, most people know these foods are highly recommended, but they may not be sure exactly why. To complicate matters, health experts are also encouraging us to eat plenty of different-colored produce, although again, the reasoning behind the recommendation may not be quite clear.

None of this advice is new. Mothers have been telling children to "eat a lot of color" for decades. But today, scientists can explain why eating a wide variety of colorful fruits and vegetables is good for health. These foods are rich sources of—among other things—health-promoting plant compounds known as phytochemicals. There are thousands of different phytochemicals in the plant world. Scientists have isolated and studied many of them, and so far, some of the most potent fall into two categories: carotenoids and flavonoids.

Carotenoids are plant pigments that have vitamin-like properties. And although their name may suggest they are only found in carrots, that's not the case at all. These colorful nutrients provide the orange, yellow, red, purple, and other colors of many fruits and vegetables. At the same time, they promote good health and reduce the risk of disease.

You may have heard of some of the better-known carotenoids, such as beta-carotene, lutein, and lycopene. During the past twenty years, scientists have conducted hundreds of studies on these important substances. Research has shown, for example, that carotenoids are powerful antioxidants, much like vitamins C and E, with the ability to minimize the damage done by rogue molecules known as free radicals, and to protect against such common ailments as heart disease, cancer, diabetes, Alzheimer's disease, chronic fatigue syndrome, and vision problems.

Carotenoids work both independently and together, enhancing one another's effectiveness. And they are remarkably versatile. Some carotenoids, for example, can be converted by our bodies to vitamin A when more of this nutrient is needed. Other carotenoids have specific disease-fighting abilities. Beta-carotene, for example, is a powerful stimulant of immune cells that protects against infections and certain types of cancer. Lutein can reduce the risk of developing macular degeneration, a leading cause of blindness. And lycopene may help prevent prostate, stomach, and lung cancer.

Flavonoids are another large family of phytochemicals. The largest subgroup of a family of plant compounds known as polyphenols, flavonoids are compounds found not only in fruits and vegetables but in tea, nuts, beans, herbs, legumes, and grains, as well. Some flavonoids, including the soybean-derived genistein and EGCG (epigallocatechin gallate) from green tea, have made headlines. With its weak resemblance to the hormone estrogen, genistein is shaping up as a useful supplement for women suffering from menopause discomforts. And EGCG is being hailed for its cancer-fighting and weight-loss qualities. But there are a number of other, equally powerful flavonoids that can benefit your health in a wide variety of ways.

With healthcare and prescription drug costs surging upward, it's encouraging to know that Mother Nature has provided us with a veritable arsenal against disease in the form of these phytochemicals. The problem is that the Standard American Diet (with the appropriate acronym *SAD*) is woefully lacking in carotenoids and flavonoids. Trade in the typical burger, with its wilted iceberg lettuce and unripened tomato slice, for a real salad and you've taken a healthy step toward an entirely new you. The more processed baked goods, fast food, and snack products you can exchange for "real" food—fruits, vegetables, grains, and nuts— the more carotenoids and flavonoids you'll be getting. Incorporating more carotenoids and flavonoids into your meals isn't difficult, because they are so abundant in nature. And once you understand the benefits bestowed on your health by these potent substances, you'll want to seek out these foods.

Although we've been led to believe that the real benefit of fruits and vegetables is in their vitamin C content, that's not really accurate. Most of the antioxidant power in these and other whole foods comes from carotenoids and flavonoids, the unsung heroes of good health. There are thousands of these compounds in fruits and vegetables, some working together, others operating in slightly different ways. And scientists are discovering that one or two antioxidants are not enough to protect us against disease. Instead, our bodies seem to thrive when we consume many different antioxidants, like those found in a diet focused on a variety of fruits, vegetables, and other whole foods.

In addition to providing your body with the building blocks of good health, there's one more area where carotenoids and flavonoids excel— weight management. No, they're not a magic bullet that will make pounds disappear. But a diet high in these substances definitely falls into the low-carbohydrate, high-fiber category. Better yet, the carbs in these foods are primarily complex, the very kind that provide the slow, steady supply of energy essential for blood sugar and hunger management.

Add some high-quality protein to the mix and you've got the makings of a solid weight-loss plan.

In the following pages, we'll look at the foremost carotenoids and flavonoids, as well as some that are considered up-and-coming, and learn more about recent scientific research showing how these substances can keep us healthy. We'll also explore both sides of the ongoing debate over food versus supplements, and examine ways to increase intake of carotenoids and flavonoids. In the end, you'll know how these vital substances enrich your health and why. More important, you'll be able to identify the everyday foods with the greatest health benefits. Whether you want to improve your overall health, treat a specific condition, or just maintain the status quo, carotenoids and flavonoids can help support your goal.

CHAPTER 1

THE POWER OF PLANTS: CAROTENOIDS AND FLAVONOIDS

Not so long ago, human beings evolved eating a diet rich in plant foods, including berries, roots, and shoots that we would probably not even recognize today. In fact, our current Standard American Diet is very limited in terms of plant foods. We concentrate on the commonplace—potatoes, iceberg lettuce, tomatoes, orange juice—and ignore anything unusual. As a result, we miss out on a wide variety of tastes and textures, as well as an incredible array of nutrients that nature provides.

All of this makes sense when you understand that carotenoids and flavonoids play an important role in nature—they provide protection for plants and animals. Sunlight is necessary for plants to grow. However, ultraviolet (UV) wavelengths can generate dangerous molecules called free radicals, which injure living cells. Free radicals oxidize cells much the same way they turn iron to rust or make butter go rancid. Many carotenoids and flavonoids function as antioxidants, which means they protect against free radicals. In other words, carotenoids and flavonoids are a plant's defense against free radicals, and they also protect plant genes, which direct the behavior of cells. And when we eat these substances, we receive the same overall health benefits. Like vitamins, carotenoids and flavonoids play many other roles in health, such as enhancing the immune system, your defense against infection; preventing damage to deoxyribonucleic acid (DNA, the molecule that carries our genetic information); and reducing the risk of cancer and heart disease.

Basically, carotenoids are pigments, the substances that make carrots orange, corn yellow, and tomatoes red, as well giving salmon, shrimp, and flamingos their vivid hues. Many flavonoids are pigments, too, although some provide flavor, as their name suggests. They are just as powerful and essential to good health as carotenoids. In this chapter, we'll learn more about these substances in general, and then look at specific carotenoids and flavonoids in subsequent chapters.

Carotenoids Defined

The term *carotenoid* (pronounced *kuh-ROT-uh-noid*) refers to a family of about 600 different pigments, a number that is likely to change as new carotenoids are identified. Despite the large number of carotenoids in nature, only a relatively small number of them play roles in human health. Around fifty carotenoids are found in the foods most Americans eat, for example, but only fourteen have been identified in the bloodstream, and just seven play significant roles in health. These seven are: alpha- and beta-carotene, astaxanthin, beta-cryptoxanthin, lutein, lycopene, and zeaxanthin.

Since science is still in the process of learning about these phytochemicals, it's too early to say whether these seven are the only ones we need. Even if all fifty commonplace carotenoids are not actually absorbed by the body, it's possible that the unabsorbed ones have some beneficial effects as they pass through the digestive tract.

Many studies have found that eating foods rich in carotenoids reduces the risk of different types of cancer, the second leading cause of death in the United States. Carotenoids may also lower the risk of heart disease, the number one cause of death. They have many other benefits, as well. For example, carotenoids have the ability to prevent our vision from deteriorating with age, reduce the risk of developing type 2 (adult-onset) diabetes and

Alzheimer's disease, increase resistance to sunburn, and prevent infertility in men.

One benefit of a carotenoid-rich diet is that many of these substances can be converted in the body to retinol, an active form of vitamin A, helping to prevent a deficiency of this vitally important nutrient. These so-called provitamin A carotenoids include alpha- and beta-carotene and beta-cryptoxanthin. A smart feature of this conversion process is that when optimum levels of vitamin A are reached, the conversion ceases, thereby avoiding toxic side effects that can occur with mega-doses of synthetic vitamin A.

There is another group of carotenoids that do not convert to vitamin A—the nonprovitamin A substances—that have additional health benefits. Researchers at Washington State University have found that these nutrients, a group that includes lutein, lycopene, and astaxanthin, not only serve as antioxidants but also stimulate the immune system. In fact, experiments have shown that the non-provitamin A carotenoids are as effective as—and at times even more effective than—the hardworking beta-carotene when it comes to strengthening immunity.

A growing body of evidence is showing that as a group, carotenoids are major players that can make a tremendous difference to health. German researchers, for example, have determined that a number of different carotenoids provide protection against colorectal cancer, one of the four most common cancers in the United States. Two other recent studies—one from New York University School of Medicine, the other from Johns Hopkins University—linked high levels of various carotenoids to a reduced risk of breast cancer. And in Japan, scientists noted that a diet rich in assorted carotenoids from fruits and vegetables protected against high blood sugar levels, a factor in the development of type 2 diabetes.

But as we will see in Chapters 2 and 3, individual carotenoids are also being tested in laboratories all over the world. Scientists are accumulating considerable evidence showing that some members of this group—astaxanthin, alpha- and beta-carotene, beta-cryptoxanthin, lutein, zeaxanthin, and lycopene, for example—are impressive weapons against many of today's most common diseases.

Flavonoids Explained

Like carotenoids, flavonoids (*FLAV-uh-noids*) are primarily found in plants. Some flavonoids are pigments, but others aren't. There are more than 6,000 compounds in the flavonoid category, and thankfully we only need to concentrate on a few that have been studied extensively and have proven health benefits. Because the flavonoid family is so large, these substances have a wide range of potency and selectivity. Certain flavonoids have been shown, for example, to reduce the risk of cancer and heart disease, while others tame menopause symptoms or strengthen bones.

Flavonoids have more in common with carotenoids than just their plant-based origins. As antioxidants, they, too, have the ability to defend the body's cells against the ravages of free radicals, the rogue molecules that can wreak havoc on health. Our bodies are designed to repair free-radical damage. But like a car mechanic who can fix an engine only if he or she has the right tools, the body can't correct free-radical damage, also known as oxidation, unless it has antioxidant "tools."

Oxidation
The process of combining with oxygen, or an oxidizing agent. Harmful changes in a cell or molecule can be caused by exposure to oxygen.

Our bodies can produce some antioxidants. Fat-soluble nutrients, such as vitamin E and the carotenoids, can be stored in the body—if they're available, of course. Water-soluble nutrients, on the other hand, are not stored, so we need to continually replenish supplies of vitamin C, for example, as well as many others, including flavonoids.

Fat Soluble
All carotenoids are fat-soluble nutrients, meaning that they can dissolve in fats and oils, but not in water.

But as with carotenoids, the benefits of flavonoids don't end with their antioxidant capabilities. Some have been shown to be effective antihistamines, helping reduce the allergy symptoms suffered by millions of people. Others fight viruses or inflammation, a condition increasingly recognized as a major factor in a host of serious health problems, including heart disease, osteoarthritis, and many others. Flavonoids also have a special synergy with vitamin C, so that each makes the other more effective.

Flavonoids, also sometimes referred to as bioflavonoids, are found throughout nature. Fruits and vegetables, of course, are good sources, as are berries, beans, legumes, herbs, tea, nuts, and grains. Some come from sources that are not normally part of our diet, such as pine bark extract, while others are primarily found in parts of food that we don't eat, like grape seeds or the white pithy substance beneath citrus peels.

The Next Best Thing to Vitamins

Currently, carotenoids and flavonoids are not considered vitamins officially, because they have not been identified as nutrients essential to preventing a specific disease. There is substantial evidence, however, that they are highly beneficial, and because our bodies cannot make these substances on their own, they must be obtained from the diet. In fact, because people consumed large quantities of carotenoids and flavonoids over millions of years of evolution, some researchers believe that people and plants may have coevolved in a way that makes us dependent on these nutrients for health.

Right now, it's hard to absolutely prove that carotenoids and flavonoids are essential to health. One reason is that researchers tend to look for fairly rapidly developing signs of nutritional deficiencies to determine whether or not a nutrient is essential. Heart disease and cancer, however, take decades to develop, making it harder for scientists to connect a deficiency of a carotenoid or flavonoid with these diseases. In the near future, many carotenoids and flavonoids could be recognized as essential nutrients, though. Meanwhile, there is plenty of evidence that they are extremely beneficial to our health.

Vitamins
Fat-soluble or water-soluble organic substances essential for the body's normal growth and functions, and obtained from plant and animal foods.

A History Lesson

Carotenoids and flavonoids are rapidly gaining respect among health experts, but they are not new by any means. Carotene was first isolated by a scientist from the roots of carrots in 1831, and a few years later another scientist identified xanthophylls in yellow autumn leaves. In 1911, yet another scientist recognized that these were related compounds, and he coined the term *carotenoids*.

Flavonoids were discovered by accident in 1938. Nobel Prize winner Dr. Albert Szent-Györgyi was treating a patient with vitamin C and concocted one particular formula that worked far better than previous versions. He discovered that the reason for this success was that the formula contained more than pure vitamin C. He named the non-vitamin-C nutrients vitamin P and noted that the two substances, C and P, were synergistic—that is, they worked better together than either did alone. Later, vitamin P was renamed *flavonoids*. Vitamin C is still often combined with flavonoids in supplements, though, to enhance each substance's benefits.

Years of research have proven that diets rich in

fruits and vegetables are associated with a relatively low risk of disease. For example, vegetarians and Seventh-Day Adventists, whose religion advocates a meatless diet, have lower rates of heart disease and cancer compared with people who eat relatively few fruits and vegetables. And it is well established that fruits and vegetables are rich sources of carotenoids and flavonoids, as well as many other important nutrients.

Of course, that doesn't mean you need to give up meat completely. What it does mean is that the meat-and-potatoes diet needs to be expanded to include more of nature's bounty. Once you understand the many advantages gained by adding fruits and vegetables to your daily meals, the salad bar will likely become a much more appealing option.

CHAPTER 2

CAROTENOIDS CLOSE UP

Now that you understand the importance of carotenoids in general, let's get to know some major members of this family, and see where they fit into the big picture of better health.

One of the Best: Beta-Carotene

One interesting benefit of carotenoids is the ability some of these nutrients have to convert to vitamin A in the digestive tract. As you may know, vitamin A is an essential nutrient, meaning that you cannot be healthy—or even live—without it. Vitamin A plays an important role in vision, reproduction, bone health, cell division, and cell differentiation, which is how a cell determines what it is going to become. The surface linings of the respiratory, urinary, and intestinal tracts, as well as the eyes, all require vitamin A to properly function. If these linings weaken, bacteria can invade the body and an infection may occur. Vitamin A is also essential for a healthy immune system.

Beta-carotene is the best-known carotenoid with the ability to convert to vitamin A, and it is sometimes referred to as provitamin A. Other major dietary carotenoids, such as lutein and lycopene, cannot be converted to vitamin A. Beta-carotene is one of the most common carotenoids found in foods, and it is the primary precursor to vitamin A. Its orangeish color accounts for the appearance of carrots, pumpkins, peaches, and sweet potatoes.

Scientists have known for decades that beta-carotene is a source of provitamin A, and that vitamin A is an essential nutrient. An early sign of low levels of vitamin A is night blindness, a condition in which the eyes do not quickly adjust to changes in light. For example, people with night blindness walk from daylight into a dark movie theater and have to wait several minutes for their eyes to adjust, so they can find a seat. Similarly, the glare of oncoming headlights can blind nighttime drivers who have night blindness. Both "preformed" vitamin A (found in meat and eggs) and beta-carotene supplements can prevent and correct night blindness, though beta-carotene works slower because the body has to convert it to vitamin A.

Night Blindness

Impaired vision in dim light and in the dark due to either an inherited condition or a vitamin A deficiency.

Night blindness is a very common condition and one that doctors routinely ignore. If you sit in the back row of a movie theater, you'll see that a lot of people have difficulty seeing when they first walk in. Their eyes also hurt when the lights are turned on. Night blindness also precedes some very serious eye diseases, such as glaucoma and retinitis pigmentosa, which can lead to complete blindness. So it's important to obtain sufficient vitamin A from foods or supplements. Beta-carotene, however, is safer than vitamin A because the conversion process in the body is slow, and there is no danger of overdosing. When the body has sufficient supplies of vitamin A, the excess beta-carotene is simply eliminated.

Beta-Carotene Fights Cancer

Night blindness is only the tip of the iceberg when it comes to health benefits of beta-carotene, though. Perhaps its most remarkable ability involves "switching on" some of the body's immune cells, something beta-carotene can do in a number of different ways. In one recent study, conducted at the Institute of Food Research in England, scientists studied how beta-carotene influenced the effectiveness of cancer-killing immune cells known as monocytes.

Monocyte
A type of immune cell that seeks out and helps destroy cancer cells and infectious microbes.

In order to seek out cancer cells, monocytes have to be able to tell them apart from normal cells. They do this with a special protein that serves as a form of "anticancer radar." When these cancer detectors spot a cancer cell, the monocyte sends a signal to other immune cells telling them to move in and destroy the aberrant cells. However, if monocytes don't have enough cancer-detecting proteins, the cancer cell escapes unnoticed and can then create more cancer cells.

This is where beta-carotene comes in. According to the study, beta-carotene increases the number of cancer-detecting proteins on monocytes. Twenty-five healthy men were given either beta-carotene supplements or a placebo (an inactive look-alike supplement) for thirty days. The men who took 15 mg of beta-carotene daily developed large numbers of cancer-detecting proteins on their monocytes, and their production of tumor necrosis factor alpha (TNF-a)—an immune system molecule that, like a heat-seeking missile, zeroes in on cancer cells and destroys them—also increased.

A number of studies at Tufts University in Massachusetts have also found that beta-carotene supplements increase the activity of the so-called natural killer (NK) cells. As their name suggests, NK cells are powerful immune forces that attack both cancer cells and virus-infected cells. The Tufts researchers found that beta-carotene supplements boosted NK cell activity, particularly in elderly men. This is important because immune function declines with age. In one recent study, elderly men who had taken 50 mg of beta-carotene every other day for twelve years had significantly greater NK cell activity.

Researchers at Loyola University Medical Center, Maywood, Illinois, recently found that beta-carotene supplements (30 mg daily) increased immune function in fifty patients with colon cancer or colon polyps. Before taking beta-carotene, the colon cancer patients had lower percentages of three different types of immune cells that play pivotal roles in preventing and fighting cancer. Beta-carotene supplements significantly increased the numbers of two of these types of cells in the colon cancer patients, and supplements slightly increased them in patients with polyps. These results are worth noting because larger numbers of these immune cells may enhance the body's ability to destroy cancer cells.

There is considerable evidence that beta-carotene can fight other types of cancer. For example, a precancerous condition called oral leukoplakia, which occurs most often in people who smoke tobacco and drink excessive amounts of alcohol, often can be treated with beta-carotene.

Oral Leukoplakia
A precancerous lesion of the mouth or throat that often leads to full-blown oral cancer if untreated.

In a number of human studies conducted at the University of Arizona, Tucson, and other research centers, physicians have found that beta-carotene supplements can reverse oral leukoplakia. Based on the studies, 50 mg daily seems to be an effective dose. Vitamin A and vitamin E can also reverse this condition.

Beta-Carotene for Breast and Prostate Cancers

There is strong evidence that diets high in fresh fruits and vegetables can reduce the risk of most, if not all, cancers. Of course, there are many beneficial nutrients in such foods, but researchers have had some success in isolating those they believe to be the most potent. For example, scientists have found that beta-carotene and other carotenoids are associated with a reduced risk of developing breast cancer.

In one recent study, researchers at Harvard and Tufts Universities studied 109 Boston-area women who had breast biopsies. The researchers found that women with higher breast concentrations of carotenoids were less likely to have breast cancer. In contrast, women with cancer had significantly lower levels of beta-carotene, lycopene, lutein, and zeaxanthin in their breast tissue. Women with the highest breast levels of beta-carotene had the lowest risk of breast cancer. Of course, this study showed an association, not cause and effect, but other studies have revealed similar trends.

For example, researchers have found that women who ate carrots or spinach more than twice weekly had a 44 percent lower risk of developing breast cancer compared with women who did not consume any of these vegetables. Women with the highest intake of beta-carotene in foods had a 36 percent lower risk of breast cancer.

In addition, several studies have found that beta-carotene may lower the risk of a precancerous condition known as cervical dysplasia. In a study at the University of Arizona Cancer Center, Tucson, researchers found that low beta-carotene levels were the only nutritional association (out of ten nutrients analyzed) with cervical dysplasia and cancer.

Two carotenoids, beta-carotene and lycopene (which we'll look at a little later in this chapter), seem to reduce the risk of prostate cancer. In a recently concluded study of some 900 male physicians at Harvard School of Public Health, researchers assessed the effects of dietary alpha-carotene, beta-carotene, and lycopene during a five-year period. While lycopene seemed to protect men over the age of sixty-five with no hereditary predisposition to prostate cancer, a diet rich in beta-carotene provided the same cancer protection for younger men.

By now you may be wondering how beta-carotene fights cancer. Researchers have found that its cancer-fighting abilities are connected to beta-carotene's antioxidant powers. As an antioxidant, beta-carotene quenches singlet oxygen and other types of free radicals. In doing so, it prevents free radicals from damaging the cells' DNA. Damaged DNA incorrectly replicates genetic instructions, which accelerates the aging of cells and increases the risk of diseases such as cancer. It is damaged DNA that creates cancer cells and instructs them to grow uncontrollably.

Singlet Oxygen
A highly reactive free radical produced by the reaction of ultraviolet light with oxygen in the skin.

In one study, Japanese researchers took human lymphocyte cells from healthy young women and then exposed the cells to x-ray radiation. Radiation creates free radicals and increases cell damage, essentially speeding up the natural process of cell damage that occurs with age. When the women took beta-carotene supplements, however, their cells became more resistant to x-ray damage. Other experiments have shown similar benefits. For example, researchers have found that cells are better able to recover from DNA damage if they have been saturated with beta-carotene.

In addition to reducing the risk of oral, cervical, breast, and prostate cancers, there is promising animal and cell-culture research suggesting that beta-carotene can decrease the likelihood of developing colon, liver, and pancreatic cancers. Plus, preliminary research shows that beta-carotene can lessen the effects of sunburn, so it's likely that this carotenoid can reduce the risk of skin cancer.

Beta-carotene supplements, in conjunction with other antioxidants, may have an even greater role in cancer prevention. In one study, researchers gave a combination of beta-carotene, vitamin E, and selenium to more than 3,000 people with esophageal dysplasia, a precancerous condition. The supplement combination reduced the risk of both esophageal cancer and death. The important thing to keep in mind is that beta-carotene plays a role in *preventing* some cancers, but not others—and that there is no evidence to indicate that it can help treat active cancers.

Heart Disease and Beta-Carotene

Beta-carotene does more than fight cancer, though. It also shows promise in reducing the risk of heart disease, the nation's number one killer, because of its ability to lower cholesterol levels.

Cholesterol
The most common type of steroid found in the body, and a necessary ingredient in a number of body functions.

Remember that cholesterol is not inherently bad; in fact, it is essential for life. It serves as the basic building block of the body's steroid hormones (estrogen, progesterone, testosterone, and others) and is also needed for the process that enables us to convert sunlight to vitamin D. Furthermore, LDL cholesterol is the body's medium for transporting fat-soluble micronutrients (such as the carotenoids beta-carotene, lycopene, lutein, and vitamin E) through the bloodstream. Part of what leads to the oxidation of LDL is its own lack of antioxidants. So consuming plenty of antioxidants—such as beta-carotene and other carotenoids—can lower LDL oxidation.

While very high cholesterol might indicate that something is wrong, it is the oxidative or free-radical damage—known as lipid peroxidation—to the low-density lipoprotein (LDL, or "bad") form of cholesterol that seems to play a key role in the development of heart disease. LDL becomes oxidized when there are either too many free radicals or not enough antioxidants in the body.

White blood cells ignore nonoxidized LDL. But they seem to sense that there is something wrong with oxidized LDL, and they attack and swallow it as they would invading bacteria. The white blood cells are then drawn to the walls of arteries, where they get stuck, and start forming cholesterol deposits known as plaque. These plaque deposits narrow and harden the arteries, and play a major role in heart attacks. One smart way to avoid heart disease is to prevent the oxidation of LDL. You can do this with beta-carotene and other antioxidants.

Although the scientific evidence is sometimes contradictory—as scientific research often is—there is compelling data showing that beta-carotene can lower cholesterol levels and reduce heart disease risk. Its benefits to the heart seem to increase when it is combined with other antioxidant nutrients.

For example, a study conducted at the University of Washington, Seattle, found that supplements of beta-carotene, vitamin E, and vitamin C in tomato juice reduced LDL oxidation in smokers. Even on its own, beta-carotene is effective. Researchers at the University of Toronto found that 20 mg of beta-carotene daily reduced lipid peroxidation in smokers, a risk factor that tends to be higher in people with heart disease, most likely because they are not consuming adequate levels of antioxidants.

There is also intriguing animal research showing that beta-carotene can reduce cholesterol. In a recent study at the University of Nebraska, Lincoln, researchers fed rabbits, a common laboratory model for studying heart disease, a diet that contained large amounts of cholesterol, peanut oil, and coconut oil. At the same time, the animals were given beta-carotene, vitamin E, or both.

Beta-carotene and vitamin E had complementary effects. Beta-carotene decreased total cholesterol and LDL levels, the size of cholesterol deposits, and the thickness of blood vessel walls, while vitamin E reduced LDL oxidation.

Other recent research supports these findings. For example, a study published in the *American Journal of Clinical Nutrition* found that diets high in beta-carotene could reduce the risk of heart disease by about one-third in women smokers. And another study found that beta-carotene stood out among carotenoids for its association with a low risk of heart disease.

The Beta-Carotene/Lung Cancer Controversy

One area of controversy involves beta-carotene and research into its effects on lung cancer. Unfortunately, the topic is rife with misinformation. Back in 1994, Finnish researchers reported the results of the Alpha-Tocopherol, Beta Carotene Cancer Prevention Study Group (ATBC). They found that beta-carotene supplementation slightly increased the risk of lung cancer in men who were smokers or asbestos workers. These findings surprised many people, because they were the opposite of what was generally believed—that is, that beta-carotene would reduce the risk of cancer.

In another study, called the Beta-Carotene and Retinol Efficacy Trial (CARET), American researchers announced in 1996 that high supplemental doses of beta-carotene and vitamin A also increased the risk of lung cancer in smokers. This news was pretty disappointing, at least on the surface.

But there is much more to the story, and it reveals how medical research can often be distorted. For example, when the results of the CARET study were announced, the researchers acknowledged that former smokers who took beta-carotene (30 mg/day) and vitamin A (25,000 IU) supplements were 20 percent *less* likely to develop lung cancer. However, they pretty much brushed aside this positive finding and concentrated on the increase in lung cancer among current smokers. The researchers also downplayed the fact that men with the highest blood levels of beta-carotene at the start of the study were 40 percent less likely to develop lung cancer.

As the researchers continued to analyze their data, they came up with some interesting new findings. The Finnish researchers involved in the ATBC study found that a combination of beta-carotene, smoking, and high alcohol consumption increased the risk of lung cancer by 35 percent. Beta-carotene, however, did *not* increase the risk of lung cancer among people who smoked less than a pack of cigarettes daily or drank little or no alcohol. In other words, beta-carotene was not harmful as long as the men did not smoke or drink excessively. And again, it was beneficial if they had stopped smoking.

This later finding, interestingly enough, was consistent with that of a study of American physicians, which found that beta-carotene supplements had no effect, good or bad, on lung cancer. This same study of American physicians did find a decrease in the risk of prostate cancer, as have other studies, indicating that a nutrient that protects against one type of cancer may not protect against a different type.

In the CARET study, men who smoked and consumed three alcoholic drinks daily had twice the risk of lung cancer if they also took high-dose beta-carotene and vitamin A supplements. However, men in this study were taking very high daily doses of beta-carotene (one-third more than in the Finnish study), plus very high doses of vitamin A for many years. High doses of vitamin A can be toxic, and given beta-carotene's ability to convert into vitamin A, the subjects may have been taking the equivalent of 60,000 IU of vitamin A. This is an extremely high dose to be taking daily for many years, and it can lead to vitamin A toxicity within a few weeks. The researchers ignored the potential problems with excessive vitamin A, however, and instead emphasized the dangers of beta-carotene.

Of course, there is still the question of why beta-carotene would increase the risk of lung cancer by even a small degree. Most probably, the answer is that both smoking and alcohol can overwhelm the body's antioxidants—in effect, oxidizing the antioxidants—and this appears to be what happened in the ATBC and CARET studies. The oxidized by-products of beta-carotene—that is, the compounds created when alcohol damages beta-carotene—increase the toxicity of alcohol. In other words, beta-carotene alone is no match for people who are heavy smokers and hard drinkers.

However, there is strong evidence that when beta-carotene is consumed in combination with other antioxidants, such as vitamin E and selenium, they form a stronger "antioxidant network" that helps the body resist free-radical damage. It's very possible that the researchers had hoped that beta-carotene (and vitamin A) would be a silver-bullet cure for lung cancer. The reality is that nutrients, in contrast to drugs, work best as a team.

Selenium
An essential trace mineral that serves as an antioxidant.

It's also important to remember that people smoking a little and drinking a little saw no increase in lung cancer risk in the ATBC study. Only people who were smoking and drinking a lot—really abusing their bodies—had problems with beta-carotene supplementation, and most would have likely had serious health problems even if they had not taken beta-carotene. Remember, also, that former smokers saw a substantial reduction in lung cancer risk when they took beta-carotene supplements. The bottom line is that beta-carotene was beneficial to all but the group with the absolutely worst lifestyle.

Synthetic versus Natural Beta-Carotene

Here's another point worth considering: These studies used synthetic beta-carotene. There are significant differences between natural and synthetic beta-carotene, as we will see in more detail in Chapter 7, and some researchers believe that the results of these studies would have been much more positive if natural beta-carotene, or a mix of carotenoids, had been used.

Another point that was downplayed in regard to beta-carotene is that it has been shown to increase lung capacity—enabling us, for example, to breathe easy whether inhaling or exhaling, and a sign of good respiratory health. In fact, a researcher at the University of California, San Francisco, analyzed a subgroup of more than 800 men from the CARET study. He found that lung capacity, a sign of healthy lungs, was strongly linked to beta-carotene consumption—even among smokers and men exposed to asbestos dust.

Beta-carotene's effect on lung function in this study was not an anomaly, either. Several studies have found that beta-carotene, sometimes in conjunction with other antioxidants, improves lung function. In a study at Cornell University, beta-carotene stood out among other nutrients for its association with normal lung function. In other research, scientists found that a combination of beta-carotene, vitamin C, and vitamin E improved several indicators of lung function in people working outdoors in Mexico City, where air pollution is exceptionally high.

There is further proof from research with asthmatics that beta-carotene benefits lung functions. People with asthma are under "oxidative stress"—that is, they have relatively high levels of free radicals in their bodies compared with antioxidants. A number of studies have found that antioxidants, such as vitamin C, are helpful in reducing asthmatic symptoms. The

Asthma
A common disorder involving chronic inflammation in the bronchial tubes, which causes them to swell, narrowing the airways and affecting breathing.

research relating to beta-carotene and asthma is preliminary, but promising. In an Israeli study, thirty-eight patients with exercise-induced asthma took either beta-carotene supplements or a placebo. Patients taking the placebo developed breathing problems after exercising. More than half of the patients taking beta-carotene, however, were protected against exercise-induced asthma.

Surviving Sunburn with Beta-Carotene

There's one other area where beta-carotene shines, and that is its ability to prevent—or at least minimize—sunburn. In a German study, researchers asked twenty young women to take either 30 mg of beta-carotene or a placebo daily for ten weeks. They were then asked, under scientifically controlled conditions, to sunbathe at various times over a thirteen-day period. While sunbathing, the women used a topical sunscreen cream. The women who had taken beta-carotene supplements and used sunscreen had less sunburn than those using a topical sunscreen alone. And another very recent study found that women who took natural beta-carotene experienced significant increases in beta-carotene stores in skin throughout their bodies.

Why would beta-carotene help minimize sunburn? Ultraviolet rays in sunlight generate large numbers of free radicals, which damage skin cells. The skin is rich in a number of antioxidants, including beta-carotene, vitamin E, and vitamin C. These antioxidants are quickly used up in neutralizing free radicals. Supplements of beta-carotene bolster the antioxidants in the skin, so the skin is better able to withstand and counteract free radicals.

Used in such a way, beta-carotene supplements may also reduce the long-term risk of skin cancer, though there are no studies that have yet proved this. We do know that sunburn increases the risk of skin cancer, and that skin cancers likely begin with free-radical damage. By reducing free-radical damage, beta-carotene should reduce the risk of skin cancer.

Skin damage from sunlight begins to take place within a couple of minutes of exposure. In addition, excessive exposure to sunlight suppresses the body's immune system, which would limit its ability to protect against cancer cells. Again, beta-carotene supplements seem to prevent sunlight-induced immune suppression.

The best approach to protect against skin cancer is to limit your exposure to sunlight to fifteen minutes or less daily—after all, a little sunlight is good for health. If you spend a lot of time outdoors in the sun, it would be prudent to take supplements of beta-carotene and perhaps other antioxidants, such as vitamins C and E. Of course, it's also smart to cover your skin to minimize skin damage from sunlight.

Overall, beta-carotene has a great deal to recommend it. As we have seen, it can enhance the activity of immune cells, improve lung function, improve night vision, and increase resistance to sunburn. When combined with other carotenoids and other antioxidant vitamins, beta-carotene also seems to reduce the risk of cancer and heart disease, too, making it a very powerful ally in the quest for good health.

All about Lutein

Lutein rivals beta-carotene as one of the most common carotenoids in the American diet. It provides the rich yellow color of corn and egg yolks. Lutein (pronounced *LOO-teen*) is also found in kale, spinach, and broccoli, but darker pigments mask its orange color in these foods. Recent studies have found that lutein is virtually essential for normal vision, and it may play important roles in preventing heart disease and cancer. Zeaxanthin (pronounced *zee-uh-ZAN-thin*) is closely related, but the body can convert some lutein to zeaxanthin. Zeaxanthin is found in okra, watercress, and chicory leaf.

In terms of chemical structure, lutein and zeax-anthin are slightly different from beta-carotene, but they are similar in that they, too, are antioxidants and fat soluble. Both nutrients are very common in the American diet. Depending on the fruits and vegetables you eat, you may actually consume more lutein than beta-carotene.

The Eyes Have It

Lutein and zeaxanthin are necessary for normal vision, and the lion's share of lutein research has focused on its role in eye health. While it is not currently recognized as an essential nutrient, which would reclassify it as a vitamin, this could happen within the next few years.

Here is how lutein is involved in vision. As light passes through the lens of the eye, it focuses the back of the eye, an area known as the retina. The retina is something like the screen in a movie theater—this is the biological screen onto which images are projected. From the retina, these light images are converted to optical signals that transmit the image through the optic nerve to the brain.

Retina
The nerve layer lining the back of the eye. Light is sensed in the retina, creating impulses that travel through the optic nerve to the brain.

At the center of the retina is a small area known as the macula, the part of the eye responsible for detailed vision. The macula contains what is known as the macular pigment, which consists primarily of lutein and zeaxanthin. So far, no other carotenoids have been identified in the macular pigment.

In addition to picking up fine detail, lutein and zeaxanthin filter out harmful wavelengths of blue light. These wavelengths can generate damaging free radicals in the eye. As antioxidants, lutein and zeaxanthin also seem to reduce the amount of free-radical damage in this part of the eye.

When the eye does not have sufficient supplies

of lutein and zeaxanthin, the macular pigment becomes thinner, so fewer harmful blue wavelengths get filtered out. At the same time, levels of drusen, a type of oxidized fat, increase. Both a thin macular pigment and an increase in drusen are indicative of macular degeneration, the leading cause of blindness among Americans over age sixty-five. Macular degeneration causes complete blindness in 300,000 Americans and a partial loss of vision in an estimated 13 million others.

Macular Degeneration

In the wet form, a condition in which fluid leaks from newly formed blood vessels, lifting the macula and distorting vision. In the dry form, there is no leakage.

Many studies strongly suggest that lutein can help in macular degeneration. People with both the "dry" and "wet" forms of macular degeneration tend to have low lutein intakes and blood levels of the nutrient, as well as thin macular pigments.

It is probably not possible to reverse advanced macular degeneration, but researchers believe that high intake of lutein supplements or lutein-rich foods will slow the progression of this disease. It's conceivable that large amounts of lutein may reverse the early stages of macular degeneration. And it does appear that lutein can prevent macular degeneration and delay its severity.

For example, there have been a number of studies showing that eating diets rich in lutein or taking 30 mg daily of lutein supplements can significantly increase the thickness of the macular pigment within about five months. In one study, the macular pigment increased by as much as 39 percent, an amount high enough to decrease the amount of harmful blue wavelengths by 40 percent.

In another study, described in the *Journal of the American Medical Association*, researchers found that people consuming the largest quantity of lutein-rich vegetables—about 6 mg of lutein daily—had the lowest risk of macular degeneration.

As for zeaxanthin, the body can convert some lutein to zeaxanthin. The most sensible approach to protect eye health is to eat a lot of carotenoid-rich foods, which are chiefly fruits and vegetables. But if you are at risk of macular degeneration, it might be worthwhile to take lutein supplements, too.

Beyond Better Vision

Research has shown that, like antioxidants, lutein reduces the oxidation of LDL cholesterol, thus minimizing the risk of heart disease. In comparing the diets of people living in Toulouse, France, and Belfast, Ireland, researchers noted that the relatively low risk of heart disease in Toulouse was linked to high blood levels of lutein and another carotenoid, beta-cryptoxanthin. In contrast, low levels of these carotenoids were associated with a higher risk of heart disease in Belfast.

Lutein and zeaxanthin may also provide protection against breast cancer, according to two experiments at Washington State University. In both trials, mice were fed a diet with different amounts of supplemental lutein, or no lutein at all. After two weeks, the mice were injected with breast cancer cells. The results of the first experiment showed that all the animals receiving lutein developed breast cancer later than did untreated animals. They also had a lower incidence of breast tumors, and their tumors were smaller than those of untreated mice. However, the lowest dose of lutein—0.002 percent of the diet—seemed to be the most effective.

The same researchers repeated the study four years later to determine how lutein was able to discourage breast cancer development. They found that the carotenoid was able to inhibit the growth of blood vessels that enable tumor growth, and also proved lethal to cancer cells without harming helpful, naturally occurring anticancer cells in the bloodstream.

All in all, lutein and its partner zeaxanthin are

probably essential nutrients for vision—the scientific community just hasn't gotten around to making it official. Considerable research on lutein and eye health shows it reducing the likelihood of macular degeneration, and some research suggests that it might protect against cataracts, as well. In the near future, we may find that lutein and zeaxanthin provide additional health benefits, but for the time being, saving our sight is too valuable to overlook.

Learning about Lycopene

Tomatoes are the richest dietary source of the carotenoid lycopene (pronounced *LIKE-o-peen*), with smaller amounts found in watermelon, pink grapefruit, guava, and dried apricots. And in addition to reducing risk of prostate cancer, lycopene fights several other types of cancer and heart disease.

Lycopene made headlines a few years ago when researchers at Harvard Medical School reported that diets high in the substance significantly reduced the risk of prostate cancer in men. The findings were based on data from a survey of almost 48,000 male physicians. Four foods were associated with a low risk of developing prostate cancer: tomato sauce (on spaghetti), pizza (with tomato sauce), raw tomatoes, and strawberries.

Furthermore, the study showed that men who ate some form of tomato (except tomato juice) two or more times a week were 21 to 34 percent less likely to develop prostate cancer, compared with men who ate few or no tomato products. Men who ate more than ten servings of tomato products weekly had a 45 percent lower risk of prostate cancer.

The greatest cancer-fighting benefits were associated with spaghetti sauce and pizza sauce, followed by raw tomatoes. There are a couple of reasons for this, which have been confirmed by subsequent research. Lycopene is normally locked into the fibrous matrix of tomatoes, which is hard for the human body to break down during digestion. The heat of cooking helps dissolve the matrix, allowing more lycopene to be released. This is one reason why spaghetti and pizza sauces actually contain more bioavailable lycopene. Another reason is that lycopene is fat soluble and requires a little fat or oil for absorption, and tomato sauces generally use some type of oil.

Raw tomatoes had some benefit, too. Although the lycopene is still largely locked into the fibrous matrix of tomatoes, raw tomatoes are usually eaten with some oil or fat, such as in salad dressing or the meat on a sandwich. Only tomato juice provided no benefits, probably because it is neither cooked nor served with fat or oil.

Ellagic Acid
A powerful antioxidant flavonoid found in many types of berries.

Strawberries are a totally different story. Despite their red color, they contain no lycopene. The benefits of strawberries may be due to another compound called ellagic acid, a very promising cancer fighter, which we will look at in Chapter 5.

The same Harvard researchers conducted two additional studies and found the exact same patterns. The more tomato sauces men consumed, the lower their risk of prostate cancer. These experts have pointed out that while lycopene is the principal carotenoid in tomatoes, there are small amounts of others, including beta-carotene and a number of little-known carotenoids, including phytoene, gamma-carotene, phytofluene, and zeta-carotene. It's possible that the entire complex of carotenoids is better than just lycopene. Still, there is compelling evidence to support the benefits of lycopene alone.

How Lycopene Works

As carotenoids go, lycopene is the most powerful antioxidant, followed by beta-carotene. As a consequence, lycopene is a very efficient quencher of free radicals and thus reduces the risk of cancer and other age-related diseases. But lycopene and other carotenoids have many non-antioxidant functions that seem to influence gene behavior. As scientists learn more about lycopene, they may discover that it turns on beneficial genes and turns off disease-causing genes.

Lycopene seems to predominate in certain organs. For example, the testes and the adrenal glands contain the body's largest stores of lycopene, compared with other organs and glands, such as the kidneys and ovaries. When organs retain larger amounts of a particular nutrient, it suggests that the nutrient plays an important role in those organs. A good example is the high level of lutein stored in the eye's macula.

Compared to other cancer-protective carotenoids, preliminary studies show that lycopene is quite powerful. In laboratory experiments, lycopene was far more effective than beta-carotene and alpha-carotene in inhibiting the growth of endometrial, breast, and lung cancer cells. As much as ten times more beta-carotene was needed to achieve the same inhibition of cancer cell growth. As for lung cancer, preliminary animal studies show that lycopene inhibited the growth of lung tumors after exposure to various cancer-causing chemicals. These results don't mean that lycopene is a cure for cancer. However, such research suggests that lycopene can help prevent some cancers.

Lycopene and Heart Health

Like most antioxidant nutrients, lycopene offers protection against heart disease. A recent study highlighted how high body levels of lycopene might reduce the risk of heart attack. Lycopene and other fat-soluble nutrients are typically stored in fat. So researchers analyzed levels of lycopene and other nutrients in fat samples taken from heart attack patients and healthy subjects in ten European countries. People with the highest lycopene levels were 48 percent less likely to suffer a heart attack.

As we saw earlier, damaging free radicals play key roles in the development of heart disease. These destructive molecules can turn normal LDL cholesterol into oxidized LDL ("bad") cholesterol, which accelerates the development of heart disease. But lycopene can lower cholesterol levels. Scientists in Israel have shown that supplemental lycopene and beta-carotene can partially inhibit the body's production of cholesterol. These substances share some common biochemical pathways in the body, and loading up on carotenoids essentially leaves less room for cholesterol.

Beta-carotene, lycopene, lutein, and zeaxanthin are hot topics in research these days. But there are plenty of other related substances that merit our attention. In the next chapter, we will take a look at some of the most promising carotenoids, and see how they can improve health.

CHAPTER 3

ALPHA-CAROTENE, ASTAXANTHIN, AND CRYPTOXANTHIN

Although beta-carotene, lutein, and lycopene are the "big three" of carotenoids, there are other important members of this family. Among them: alpha-carotene, astaxanthin, and beta-cryptoxanthin, also known as b-cryptoxanthin. In this chapter, we will look at some of the ways these lesser-known substances contribute to health.

All about Alpha-Carotene

As you learned in Chapter 1, alpha-carotene is closely related to beta-carotene. And like beta-carotene, it has considerable health benefits, especially when it comes to cancer prevention.

Serious research on alpha-carotene began in Japan during the late 1980s. Researchers there were motivated by the growing number of studies on beta-carotene and believed that other carotenoids might also be worth investigating. They discovered that alpha-carotene outperformed beta-carotene when it came to anticancer properties in cell-culture experiments. Alpha-carotene was more effective than beta-carotene at delaying or inhibiting the growth of various types of human cancer cells, including those of the brain, pancreas, and stomach. Of course, cell-culture experiments are far from studies in human beings, but they do indicate the potential benefits.

One of the best human studies on alpha-carotene looked at the eating habits of male smokers in New Jersey. Some of the men had been diagnosed with cancer of the trachea, bronchus, or lung, and their diets were compared with those of comparably aged healthy men. Men consuming food low in alpha-carotene had the greatest risk of these cancers.

More recently, a handful of studies have shown that alpha-carotene—sometimes independently and sometimes with other carotenoids—has the ability to reduce high blood pressure, prevent heart disease in women, and fight ovarian cancer. For example, researchers at Tulane University School of Medicine in New Orleans found that alpha-carotene was somewhat more likely to be associated with moderate blood pressure than beta-carotene.

Meanwhile, scientists analyzing data from the Nurses' Health Study, following more than 73,000 women for twelve years, found a 26 percent reduced risk of coronary artery disease among those with the highest intake of alpha- and beta-carotene-rich foods. Previous clinical trials examining only beta-carotene's effect on heart disease had found no association, leading the researchers to speculate that alpha-carotene may be responsible for the protective effect.

Finally, scientists at Brigham and Women's Hospital in Boston found that high levels of alpha-carotene from food and supplements were closely associated with a reduced risk of ovarian cancer among postmenopausal women in a study of more than 1,000 women. Additionally, they noted that high levels of lycopene were associated with less ovarian cancer in premenopausal women. Raw carrots and tomato sauce were the foods most associated with a protective effect.

Get Acquainted with Astaxanthin

Astaxanthin (pronounced *as-ta-ZAN-thin*) is a carotenoid that serves as a common pigment in salmon, trout, algae, and other sea creatures. Actually, carotenoids function as pigments and antioxidants in many fish, and some, like astaxanthin, are added to the feed of "farm-raised" fish to enhance their color and sales appeal, as well as to poultry feed to enrich the yellow color of egg yolks.

Astaxanthin has been tested extensively as a supplement for animals being raised commercially, and has been shown to have positive effects on various aspects of health. Not surprisingly, recent research has shown that astaxanthin is a powerful antioxidant. By various measures, it scored as high as vitamin E—or nearly so—as well as outperforming all other carotenoids, including beta-carotene.

In the laboratory, animal studies have shown that astaxanthin's antioxidant abilities produce benefits that very likely extend to humans, including cancer protection, reduction in the risk of heart disease, prevention of diabetes, and combating the ulcer-causing bacterium *Helicobacter pylori*.

Helicobacter pylori
A bacterium that causes stomach inflammation and ulcers. Usually acquired from contaminated food or water, it is the most common cause of ulcers worldwide.

For example, researchers at Washington State University tested astaxanthin, beta-carotene, and another carotenoid, canthaxanthin, against mammary tumors. All three carotenoids inhibited the growth of tumors in animals, but astaxanthin performed best. Similarly, at the University of Minnesota, astaxanthin was found to suppress cancer cell growth and stimulate immune response against cancer.

Heart health, too, appears to benefit from astaxanthin, according to research in Japan. Scientists there found that astaxanthin prevented neutral LDL cholesterol from turning bad by inhibiting oxidation, the process that contributes to an increased risk of heart disease. Meanwhile, a different team of Japanese researchers determined that astaxanthin protected animals with diabetes from damage to the insulin-producing cells that typically occurs with the disease.

Last but not least, Danish scientists gauged astaxanthin's effect on *H. pylori*, the bacterium responsible for ulcers, as well as stomach cancer and other gastric ailments. They found that the number of bacteria decreased, as did stomach inflammation caused by the bacteria. Experts estimate that about half the world's population is affected by *H. pylori*, and antibiotics are generally prescribed. If these findings are duplicated, they could form the basis for a better method of treating this all-too-common infection.

Cryptoxanthin Comes to the Fore

Cryptoxanthin (pronounced *krip-to-ZAN-thin*) is another carotenoid that is not yet a household word, but still has much to recommend it in terms of health benefits, especially when it comes to reducing cancer risk. Actually, the term *cryptoxanthin* refers to a substance composed of two related molecules, beta- and alpha-cryptoxanthin, which is found in papaya, peach, pumpkin, corn, red bell peppers, and citrus fruits such as tangerines and oranges. Cryptoxanthin is second only to beta-carotene in terms of dietary carotenoids that convert to vitamin A.

Recent research has focused primarily on the beta form of cryptoxanthin, and the findings are impressive. In Chapter 2, for example, we discussed a major review of previous studies involving nearly 400,000 people and thirteen prestigious health institutions to determine the role of carotenoids in lung cancer. These experts not only cleared beta-carotene of any link to lung cancer in smokers, nonsmokers, and former smokers, but also recommended foods high in beta-cryptoxanthin as a means of reducing the risk of the disease. Those findings were echoed in a study from Singapore, where scientists examined the diet and health of more than 63,000 men and women over an eight-year time span. They found that those who ate the most beta-cryptoxanthin were 27 percent less likely to develop lung cancer.

Esophagus
A long tube connecting the throat with the stomach.

Meanwhile, researchers at Boston's Tufts University found that vitamin A and beta-cryptoxanthin provided significant protection against cancer of the esophagus.

Scientists in Japan may have uncovered the mechanism that makes beta-cryptoxanthin such an effective weapon against cancer. Extensive studies there showed that beta-cryptoxanthin stimulates the expression of anticancer genes within the body.

Cancer is not the only disease that beta-cryptoxanthin can help prevent, as other research shows. Scientists at the Mayo Clinic, for example, examined the relationship between antioxidants and rheumatoid arthritis in a group of almost 30,000 older women. Only beta-cryptoxanthin and the mineral zinc were found to provide protection against the disease. And in a new twenty-three-year-long study from Finland, researchers reported that among antioxidants, only vitamin E and beta-cryptoxanthin proved effective at reducing the risk of type 2 diabetes.

Why Mixed Carotenoids May Be Best

There are many other "minor" carotenoids in fruits and vegetables. Although everyone can enrich their health with these substances, people who smoke and/or drink alcohol regularly should make a special effort to include carotenoids in their diet. The best way to do that is by eating many different kinds of fruits and vegetables. It's impossible to put every beneficial carotenoid in a tablet or capsule. Carotenoid supplements should be used to augment a good diet, not replace it.

There's another advantage to obtaining carotenoids from food. In nature, and in the fruits and vegetables you eat, there is a natural mix of various carotenoids. While one carotenoid may predominate over another, there is always a multitude of these nutrients in foods. If you eat a diverse selection of fruits and vegetables and other plant foods (such as herbs, spices, and greens), you will consume a broad mix of carotenoids. In general, the more varied the carotenoid mix, the more these nutrients probably contribute to your overall health and resistance to disease. This is an important point, because a mix of carotenoids tends to have a synergistic or complementary effect.

Carotenoid synergy can be seen at work in any number of studies. For example, in one clinical trial, researchers asked twenty-two smokers and nonsmokers to eat fruits and vegetables high in beta-carotene, lycopene, and lutein. The foods were carrots (for beta-carotene), pear tomatoes (for lycopene), and French beans, cabbage, and/or spinach (for lutein). Together, these foods added about 30 mg of mixed carotenoids daily to the diet. After two weeks on a high-carotenoid diet, blood levels of carotenoids increased by 23 percent in smokers and 11 percent in nonsmokers. This was significant because smokers typically have lower carotenoid levels. In addition, the ability of LDL cholesterol to resist oxidation that converts it to "bad" cholesterol increased by 14 percent in smokers and by 28 percent in nonsmokers. This resistance to LDL oxidation should lower the risk of coronary heart disease.

In another study, researchers placed nine healthy women on a low-carotene diet for four months. Five of the women were given a low-dose beta-carotene supplement for sixty days, while four received placebos. Afterward, all the women were given the low-dose beta-carotene supplement for two months. During the last month, the women were also given a mixed-carotenoid supplement, containing a higher dose of beta-carotene, plus alpha-carotene, cryptoxanthin, lutein, zeaxanthin, and lycopene—all of which are found in fruits and vegetables.

The low-carotene diet reduced the activity of the

women's white blood cells. Low doses of beta-carotene were not sufficient by themselves to improve the activity of the white blood cells. However, the mixed-carotenoid supplement, containing higher amounts of beta-carotene and related carotenoids, restored normal lymphocyte activity. In effect, a mixed-carotenoid supplement compensated for a diet lacking in fruits, vegetables, and juices.

White Blood Cells
Infection-fighting cells made in the body, also known as leukocytes. Lymphocytes and neutrophils are the two most common types.

Finally, a thirteen-year-long study at the University of North Carolina, Chapel Hill, tracked the health of almost 2,000 men with elevated cholesterol levels. Experts compared the intake of total carotenoids with the risk of heart attack among these men. Overall, men eating the most carotenoids had a 36 percent lower risk of heart attack. Nonsmoking men fared even better; those who ate the greatest amounts of carotenoids had a 72 percent lower risk of heart attack.

Carotenoids are found in some of the most delicious fruits and vegetables around, including apricots, peaches, cantaloupes, red peppers, and sweet potatoes. Make an effort to include more of these and other colorful produce in your diet on a daily basis. As we have seen, there are simply too many reasons to not provide your body with these vital nutrients.

CHAPTER 4

FOCUS ON FLAVONOIDS

Now that you've seen how powerful carotenoids can be, let's take a look at a few of the best-known flavonoids. Sometimes referred to as "vitamin P," flavonoids are water-soluble plant pigments. There are various ways of categorizing them, based on their chemical structures, but scientists don't agree on any one classification system. Fortunately, we don't need to sort out these complicated systems to realize the benefits flavonoids have to offer.

Flavonoids are perhaps best known for their connection to vitamin C, since the two have a strong synergistic relationship. But flavonoids have much more to offer. Some of these substances have significant cancer-fighting abilities, for example, while others strengthen blood vessels, or combat inflammation, free radicals, viruses, bacteria, or allergic reactions. Heart disease and diabetes risk can be altered by certain flavonoids, and menopause symptoms are eased by others. These nutrients also encourage the formation of collagen in connective tissue, boost the immune system, and stimulate wound healing. Bruising is one symptom of low levels of certain flavonoids, and night leg cramps, aches, and weakness are some others.

Clearly, even though flavonoids have not been declared essential nutrients, they play an important role in good health. Because there are literally thousands of known flavonoids, and very likely others as yet undiscovered, we will focus on those that have been most thoroughly researched and have important health benefits. Even though many flavonoids are closely related in terms of chemical structure, their effects on health can vary. Don't be put off by their strange-sounding names. As you'll see, there are plenty of reasons to get to know these nutrients.

Get to Know Green Tea

No doubt you've heard that green tea is good for you. One reason it is so highly recommended is because it contains a flavonoid called epigallo-catechin gallate, mercifully abbreviated to EGCG. Actually, green tea contains quite a number of flavonoids, but EGCG is garnering a great deal of attention due to its cancer-fighting qualities. A growing number of international studies show that EGCG, and related compounds from green tea, can inhibit growth of various types of cancer cells. For instance, researchers in Poland tested EGCG against three different types of cancer cells—laryngeal, colon, and cervical—and found that in test tubes, the compound was lethal to cancer cells without damaging normal cells.

Green Tea
Made from the leaves of Camellia sinensis, tea plant, green tea is produced by steaming or roasting the leaves, then rolling and drying them.

EGCG and two related substances were pitted against breast cancer cells by New Zealand researchers, with similar results. The green tea derivatives reduced the number of cancer cells in test tubes, and decreased actual breast tumor growth in an animal study. Other research into EGCG and breast cancer has shown that the substance has a synergistic relationship with tamoxifen (an anti-cancer drug used to block the effects of estrogen), each enhancing the other's ability to kill breast cancer cells. And another animal study at the University of Wisconsin found that oral doses of EGCG amounting to approximately six cups of green tea daily significantly inhibited prostate cancer development and spread of the disease.

EGCG Equals Good Health

Several recent studies have linked EGCG to reduced risk of heart disease and high blood pressure. To determine why the flavonoid is effective, a team of researchers from South Korea and Germany examined EGCG's effect on the smooth muscle cells lining blood vessels. They found that EGCG prevents the cells from hypertrophy.

Hypertrophy *Swelling or enlargement of an organ or body part due to the increased size of its cells.*

When hypertrophy swells the lining of blood vessels, there is less space for blood flow and a greater likelihood of a clot becoming caught, blocking blood flow altogether. As the study authors noted, hypertrophy is a critical event in the development of heart disease and high blood pressure. EGCG's ability to prevent it may be why epidemiological (population) studies show that people who drink five or more cups of green tea daily not only live longer, but are also less likely to suffer from heart disease and cancer.

Finally, we should not overlook EGCG's antioxidant capabilities, which have been shown to be 164 times greater than those of vitamin C when it comes to quenching specific oxygen-related free radicals. A recent animal study from Germany provides a dramatic example of just how powerful EGCG can be at protecting us against free-radical damage. Researchers cut off blood flow to muscle tissue, a process that creates destructive free radicals. When this occurs in the body—during a heart attack or stroke, for example—cells that are deprived of blood die. At the same time, free radicals are created and the heart muscle or brain is weakened by damage from these unstable molecules. In the experiment, the muscle tissue that was treated with EGCG had 50 percent fewer free radicals than untreated tissue. In other words, EGCG protected the cells from harm by significantly eliminating free radicals.

The Soy-Health Connection

At first, Americans were slow to embrace soy foods, but reports of the health benefits inherent in soy, plus the ever-growing variety of foods created from the humble bean, are changing that rapidly. According to the latest statistics, retail sales of soy foods have grown by more than 10 percent annually for seven years in a row, topping out at $3.65 billion in 2002. Those figures have put soy foods into mainstream supermarkets, where many health experts say they belong.

Part of the reason soy foods are becoming more popular may be due to the fact that in 1999 the Food and Drug Administration (FDA) ruled that foods containing soy protein could claim to lower the risk of heart disease by raising levels of beneficial HDL cholesterol and reducing levels of LDL ("bad") cholesterol.

The FDA based its decision to endorse soy on numerous studies showing that its flavonoids, known as isoflavones, improved overall cholesterol and triglyceride profiles. For example, Dutch scientists recently compared diets of more than 900 postmenopausal women with their triglyceride levels.

Triglycerides
A form of fat, triglycerides are also the backbone of many types of other fats. They come from food and can be produced by the body.

High triglyceride levels are linked to both diabetes and heart disease, but the women who consumed the most isoflavones (genistein and related compounds) had the lowest triglyceride levels.

Similarly, researchers in Finland collected and reviewed previous studies looking for a link between phytoestrogens and heart health, and concluded that individuals with high cholesterol levels benefited more than those with normal cholesterol readings. In addition, they observed that phytoestrogens also had beneficial effects on the lining of arteries, again providing protection against heart disease. Finally, this team also noted that genistein and daidzein, a related phytoestrogen, made normal LDL cholesterol less susceptible to oxidation, the process that turns it into "bad" cholesterol.

Phytoestrogens: Natural Disease Fighters

Another likely reason soy sales are soaring has to do with the fact that its isoflavones are considered phytoestrogens.

Phytoestrogens
Plant compounds that are chemically similar to the hormone estrogen.

Phytoestrogens can match up with receptor sites on cells, preventing real estrogen from making a connection. Thus, excess estrogen is removed from the body before it has a chance to contribute to the development of breast or uterine cancer. Phytoestrogens also play an important role when menopause occurs and estrogen levels fall. In this case, phytoestrogens can once again connect with receptor sites, mimicking the effect of real estrogen.

In a comprehensive review of prior studies examining the effectiveness of genistein in treating menopausal symptoms, Italian scientists summed up the benefits of this phytoestrogen. They noted that 50 mg of genistein daily reduced the number of hot flashes, while total cholesterol and LDL cholesterol were also lowered. In addition, taking 90 mg of isoflavones for six months increased bone-mineral density, thereby protecting against the bone-thinning disease osteoporosis. There was also a possible reduction in breast cancer risk. In conclusion, the experts recommended soy extracts, and genistein in particular, to treat both short-term and long-term effects of menopause. Other health authorities have also recommended genistein and other phytoestrogens as an alternative to synthetic hormone replacement therapy (HRT), especially since HRT was recently found to be linked to an increased risk of cancer.

Treating More than Menopause

Actually, there is a considerable body of evidence showing that phytoestrogens, and genistein in particular, provide protection against cancer, especially hormone-related cancers such as breast and prostate. Population studies in Asia, where soy foods are a dietary staple, have revealed very low rates of these types of cancers compared to the United States. Further proof of the effectiveness of soy isoflavones as cancer fighters comes from studies showing that when Asians move to the United States and adopt the traditional American diet, their cancer rates soar. As a result, many researchers agree that isoflavone-rich soy foods are providing protection against cancer. Not surprisingly, genistein, the most common isoflavone in soy, is credited with many of the benefits.

One important study supporting this theory comes from researchers at the University of South Florida, who divided a group of sixty-eight pre-menopausal women in half. One segment was given 40 mg of genistein daily, while the other portion received a placebo for three months. At the end of the study, researchers determined that the women in the genistein group experienced beneficial changes in hormone concentrations and menstrual cycle length that reduced their risk of developing breast cancer.

Another clinical trial, conducted at the University of North Carolina, Chapel Hill, tested genistein and daidzein against two types of ovarian cancer cells. Both proved effective at reducing production of cancer cells by 20 percent in test tubes, and researchers noted that the amounts of isoflavones used were roughly the same as would be obtained through a diet that incorporated soy foods.

Soy isoflavones are versatile enough to help protect men against prostate cancer, another hormone-related form of the disease, according to the findings of three recent clinical trials. The first, conducted by scientists at the University of California at San Francisco, examined the diets and prostate cancer incidence in a group of almost 400 men. The data showed that those who ate the most soy foods containing the isoflavones genistein and daidzein had the lowest risk of prostate cancer.

In the second study, animals were given isoflavone supplements and then challenged with carcinogens. The supplement group had a significantly lower incidence of prostate cancer than the group that did not eat soy, leading the researchers to conclude that isoflavones could be a promising treatment for the prevention of human prostate cancer. Those findings were supported by another recent study from Japan, indicating that men who had the highest blood levels of genistein, daidzein, and equol, another isoflavone, had less risk of developing prostate cancer than men with the lowest levels.

Carcinogen *A substance that causes cancer by damaging a cell's DNA.*

Hormone-dependent cancers are not the only ones genistein and other isoflavones fight. Two new studies show that genistein is a potent weapon against pancreatic cancer. About 30,000 men and women are diagnosed with pancreatic cancer in the United States each year, a disease with a dismal 4 percent five-year survival rate.

Two recent studies, however, show that genistein may offer hope. First, researchers at Harvard Medical School found that genistein performed as well as the cancer drug herbimycin at inhibiting the growth of three different types of human pancreatic cancer cell lines.

The second study, conducted at UCLA, also found that genistein inhibited pancreatic tumor growth in animals that had been implanted with human tumors. Researchers concluded that genistein "may well be beneficial to patients with pancreatic carcinoma."

How do isoflavones such as genistein fight cancer? To answer that question, scientists at Wayne State University School of Medicine in Detroit reviewed previous studies and concluded that genistein inhibits cancer growth by regulating the genes that are involved in each cell's development and eventual death. They also observed that genistein has antioxidant abilities, and has been shown to be a potent inhibitor of new blood vessel formation in tumors and metastasis, the process by which tumors spread throughout the body.

The Versatile Genistein

Additionally, other new research is showing that isoflavones, and genistein in particular, have qualities that are only just beginning to be explored. For instance, scientists at the U.S. Department of Agriculture and George Washington University's Department of Medicine reviewed a number of previous studies involving phytoestrogens, and concluded that isoflavones and lignan-rich flaxseed play beneficial roles in obesity and diabetes. Soy protein's isoflavones, they noted, have been shown to help lower insulin, insulin resistance, blood sugar, and even body weight.

Lignan
A fiberlike substance with health-promoting properties, lignan is found in flaxseeds and wood.

Here are some other examples of genistein's health benefits. A randomized, double-blind, placebo-controlled study in Ireland found that 54 mg of genistein daily was as effective as synthetic hormone replacement therapy in preventing menopause-related bone loss, without the undesirable side effects of pharmaceuticals. And in England another double-blind, randomized, placebo-controlled clinical trial tested isoflavones (genistein, daidzein, and biochanin A) derived from red clover on middle-aged women to determine the compound's effect on bone density. When compared to women who had taken a placebo, the women in the isoflavone group had significantly less bone loss in their spines, again without experiencing any negative side effects.

Meanwhile, Swedish scientists have found that genistein has potent antibacterial properties when it comes to fighting off various types of staphylococcal strains, common and dangerous bacteria, and was also found to inhibit growth of *Helicobacter pylori*, the bacterium responsible for ulcers.

Finally, research in the People's Republic of China has shown that genistein protects the brain's neurons from damage by beta-amyloid, a protein associated with Alzheimer's disease that contributes to thick deposits of plaque implicated in brain cell death.

Does Soy Have a Downside?

Researchers are still trying to sort out whether or not the estrogenic effects of soy's isoflavones are a double-edged sword. For example, one concern involves different results in studies testing soy and breast health. While some research shows that soy inhibits breast cancer cell growth, others have shown that breast tissue growth may be stimulated by soy. Is this a risk factor for cancer? Experts say they can't answer that question without further studies.

Another area of concern comes from research showing that genistein can stimulate estrogen-receptor-positive (ER+) breast cancer growth and interfere with the antitumor activity of tamoxifen, a drug commonly prescribed to breast cancer patients. On the other hand, genistein seems to inhibit estrogen-receptor-negative (ER–) breast cancer cell growth, so in these cases consuming soy and isoflavones make sense. Anyone who is concerned about breast cancer or who has been diagnosed with the disease should discuss these issues with a nutritional expert, and determine the best individual solution.

Meanwhile, many health authorities feel that the benefits of soy may be available only to those who are raised on a steady diet of soy foods from infancy. Still, they say, starting to eat soy later in life should be safe for healthy individuals. Soy's estrogenic effects are reportedly only 1/400th to 1/1,000th as powerful as the human hormone estrogen, and its other health benefits are too numerous to ignore.

Quercetin Conquers Illness

"An apple a day keeps the doctor away" is good advice, according to a nearly thirty-year-long study of more than 10,000 people in Finland. Those who ate the highest intake of quercetin (apples and onions are good sources) had the lowest risk of dying from heart disease, as well as a reduced risk of developing lung cancer, asthma, and type 2 diabetes.

Quercetin also reduces the risk of heart disease by minimizing hardening of the arteries, lowering high cholesterol levels, and strengthening blood vessels. Researchers in Spain also found from animal studies that quercetin reduces high blood pressure and helps dilate blood vessels, preventing blockages associated with heart attacks. Furthermore, British scientists recently reported research confirming previous studies that had shown quercetin's role in preventing blood platelets from clustering, a process that can lead to clot formation. And a study involving nearly 5,000 men and women in the Netherlands found that high dietary intake of quercetin and two other flavonoids from black tea—kaempferol and myricetin—were significantly associated with a reduced risk of death from heart attack.

Platelets
Irregular, disc-shaped bodies in the blood that are involved in blood clotting.

Known primarily for its antihistamine and anti-inflammatory properties—quercetin was found to lower the release of histamines by as much as 96 percent in a recent Japanese study—this flavonoid actually provides a long list of health benefits. People with diabetes may be able to use quercetin to regulate blood sugar levels and protect their eyes, kidneys, and nerve cells from damage associated with the disease. In addition, a new study from India shows that quercetin eases the neuropathic pain caused by disease or malfunction of the nerves (neuropathy) commonly experienced by people with diabetes.

Quercetin is also a potent virus and cancer fighter, and has been shown to reverse mental dysfunction in older animals, as well as protecting brain cells from damage by strokes. Quercetin's ability to reverse age-related mental impairment was discovered by researchers in India, who noted that thirty-day treatment with the compound also elevated levels of brain antioxidants in elderly lab animals. Meanwhile, Italian scientists established the link between quercetin and several other flavonoids and decreases in brain damage during strokes.

When it comes to controlling cancer, quercetin is in a class by itself, as researchers at the State University of New York at Buffalo recently found. They tested quercetin against highly aggressive and moderately aggressive prostate cancer cell lines. Not only did quercetin significantly inhibit the cancer cell growth, but it also stimulated the production of several genes that suppress tumors.

Foods rich in quercetin include onions, apples, kale, green tea, red cabbage, tomatoes, green beans, lettuce, grapes, and potatoes.

Proanthocyanidins Promote Good Health

It's easy to be intimidated by long names that sound like something from a chemistry textbook. But it would be a shame to avoid these vitally important nutrients, because they are outstanding antioxidants with significant health benefits. Before looking at the attributes of proanthocyanidins, we should address one confusing aspect: the differences (and similarities) between two major players in this group—Pycnogenol and grape seed extract.

Pycnogenol and Grape Seed Extract

Technically, these two substances are both proanthocyanidins. The first one to reach the market was Pycnogenol, a patented, proprietary formulation derived from the bark of French maritime pine trees (*Pinus maritima*). Extracts of pine bark are actually plant pigments, similar to those found in tea, bee pollen, beans, and grapes and grape seed extract, which some experts say is comparable to pine bark in terms of health benefits. Grape seed extract, taken from the seeds (and sometimes skin) of grapes used in wine making (*Vitis vinifera*), is an excellent source of several dozen different antioxidants. Are they interchangeable? Health experts are divided on that question. After reading a sample of studies on both, you may want to form your own opinion about which one is right for you. Let's look at Pycnogenol first.

Pycnogenol can benefit health at any stage of life, but it's especially helpful when it comes to reducing the effects of aging. For example, a number of clinical trials have shown that Pycnogenol fights heart disease and stroke by thwarting the formation of artery-clogging blood clots, keeping blood vessels flexible and strong, and by reducing blood pressure. It also minimizes symptoms of asthma and inflammation, enhances the activity of other antioxidants such as vitamins C and E, and retards photoaging of the skin due to sun exposure. As two of the few substances that are capable of crossing the blood–brain barrier, grape seed extract and Pycnogenol are powerful allies when it comes to protecting the brain from free radicals. Both are also recommended for immune system health, as well as reducing inflammation and allergies.

In medical journals, there are plenty of compelling studies documenting Pycnogenol's effectiveness. At the University of California's Laboratory for Atherosclerosis and Metabolic Research, for example, researchers measured the changes in healthy subjects' antioxidant and cholesterol levels after taking 150 mg of Pycnogenol for six weeks. They found the supplements significantly increased antioxidants in the blood. Even better, levels of LDL ("bad") cholesterol dropped substantially, and HDL ("good") cholesterol increased.

At Loma Linda University's School of Medicine, another new study demonstrated Pycnogenol's ability to prevent the brain cell death that occurs with Alzheimer's disease. And in Germany, scientists discovered that 100–125 mg of Pycnogenol performed as well as 500 mg of aspirin as a method of lowering blood pressure in smokers, without causing stomach bleeding. Finally, a review of five clinical trials testing Pycnogenol as an effective treatment for diabetic retinopathy concluded the supplements produced a favorable outcome in the majority of patients.

Retinopathy
A disease of the retina, the light-sensitive membrane at the back of the eye. It may be caused by hardening of the arteries, high blood pressure, or diabetes.

The Goodness in Grape Seed Extract

For centuries, grapes and grape products have been favorites all over the world, as both food and medicine. In the nineteenth century, grapes were used to cure high blood pressure as well as treating varicose veins, diarrhea, and other ailments. Recently, compounds that prevent blood platelets from becoming "sticky," and therefore more likely to form clots, were discovered in red grape juice, making it a good alternative to wine when it comes to overall heart health.

Grape seed extract, too, is useful for reducing

heart disease risk. Its antioxidants protect LDL ("bad") cholesterol against oxidation, and shield the lining of the blood vessel walls. When fed to rabbits or mice in clinical trials, grape seed extract significantly improved heart function, inhibited the development of atherosclerotic lesions, and reduced heart tissue damage. In a human clinical trial, grape seed extract supplements significantly reduced LDL oxidation in subjects with elevated cholesterol levels.

In addition to protecting the heart, grape seed extract helps fight cancer, as two recent studies from the University of Colorado Health Sciences Center show. In one, researchers found that grape seed extract prevented angiogenesis.

Angiogenesis
The process of new blood vessel development that allows tumors to grow.

In the second study, a patented, proprietary formulation of grape seed extract was tested on human prostate cancer cells. The grape seed extract strongly inhibited tumor growth and cancer cell reproduction, while stimulating secretion of substances that diminish cancer expansion.

Additional research has shown that grape seed extract strengthens the tiny blood vessels known as capillaries, and can be used to treat or prevent circulatory disorders, such as peripheral vascular disease. Furthermore, a recent Japanese study showed that grape seed extract prevented the development of cataracts in animals genetically predisposed to the disorder. And an intriguing study at Rutgers found that grape seed extract may be helpful for weight loss. Researchers found that the extract inhibited two fat-metabolizing enzymes, suggesting that it could limit dietary fat absorption and fat deposits in the body. Although this study needs to be duplicated and confirmed, it shows promise in a field where there is little help available.

While grapes may contain some proanthocyanidins—as do cranberries, blueberries, almonds, peanuts, cocoa, and some other nuts and berries—the best-known sources are wine grape seeds and bark of the French maritime pine tree. So although most of the substances discussed in this book are ideally obtained through diet, Pycnogenol and grape seed extract are two exceptions. In these cases, supplements are the preferred form, something we'll look at in more depth in Chapter 7.

As you can see, these members of the flavonoid family have a great deal to offer when it comes to promoting good health. But there are some lesser-known flavonoids that are just as important. In the next chapter, we will learn more about them.

CHAPTER 5

MORE FLAVONOID FAVORITES

With more than 5,000 flavonoids identified, it is not realistic to include information on all of them. But here is an overview of a few other flavonoids that have demonstrated significant health-promoting abilities.

Citrus Flavonoids

Research has shown that citrus flavonoids are some of the most powerful natural cancer fighters available. Furthermore, they provide other benefits, such as preventing bone loss, lowering cholesterol, and reducing inflammation, to name only a few. There are quite a few different citrus flavonoids. Hesperidin and naringen are two prominent members of this family. Lemons and oranges are the most common sources of hesperidin, and naringen is found in grapefruit.

In recent research focusing on hesperidin and a related compound, hesperetin—both found in orange juice—scientists have found it to be a potent cancer fighter. Several animal studies with these compounds have shown that they can prevent the development of cancers of the breast and colon. And scientists at the U.S. Horticultural Research Laboratory in Florida found that hesperetin inhibited production of enzymes that transform toxins such as pesticides and cigarette smoke into cancer-causing agents known as carcinogens. Furthermore, Japanese researchers found that hesperidin prevented bone loss in an animal model of menopause. Hesperidin was also associated with increased bone absorption of strengthening minerals, such as calcium, zinc, and phosphorus.

In a recent Brazilian study, naringenin, abundant in grapefruit, performed well as a cholesterol-lowering agent (as did rutin, another flavonoid we will look at a little later in this chapter). Meanwhile, a related compound known as naringenin is gaining renown for its role in brain health. British researchers found that naringenin (and hesperetin) were the most effective flavonoids tested at crossing the blood–brain barrier. Furthermore, in an animal study, scientists in Korea noted that naringenin derived from a type of orange found in Asia inhibited the development of acetylcholinesterase, an enzyme linked to the development of Alzheimer's disease. In addition, naringenin also provided significant protection for the brain from a chemical that induces amnesia. And yet another recent animal study from Korea found that two forms of naringen were able to prevent the formation of lesions in blood vessels that contribute to hardening of the arteries and heart disease.

More Citrus, Better Health

These are only a few of the many health-promoting flavonoids found in citrus. Analyses of orange juice have shown that it contains as many as sixty such compounds! Science is only beginning to explore the health potential of some of these lesser-known substances. For example, limonin and nomilin, members of the limonoid group of citrus flavonoids, have been shown to prevent the duplication of human immunodeficiency virus-1 (HIV-1) in human blood cells.

Limonin
The white, bitter, crystalline substance found in orange and lemon seeds.

Other citrus flavonoids, such as limonene, luteolin, and diosmin—all found in lemons and oranges—have been shown to inhibit the growth of cancer in various ways. Tangeretin, a related compound from lemons, grapefruit, and oranges, not only fights cancer but also supports the part of the brain involved in Parkinson's disease. And nobiletin, from tangerines, minimizes production of an inflammation-stimulating substance, and appears to slow the progression of joint damage related to osteoarthritis.

The best way to get plenty of these nutrients is to eat a variety of citrus fruits. When Australian scientists analyzed the data from almost fifty international studies on the health benefits of citrus, they found plenty of reasons to include oranges, grapefruits, and tangerines in the daily diet. For example:

- There is solid evidence that obesity, diabetes, and heart disease can be reduced with daily citrus consumption.

- The likelihood of having a stroke declines by close to 20 percent among people who eat five daily servings of fruit and vegetables plus one extra citrus fruit.

- Stomach, larynx, mouth, and esophageal cancer rates plummet by as much as 50 percent among people who eat the most citrus fruits.

Of course, flavonoids can't take all the credit. Citrus fruits are filled with antioxidants and fiber, too, plus there are undoubtedly other substances in these foods that have yet to be discovered.

A daily glass of orange juice is a good start, but juice is high in sugar and low in fiber. A better way to get citrus flavonoids is by eating citrus fruits. Even squeezing lemon or lime wedges into water or tea is likely to provide some flavonoids, since many of these nutrients are found in the white, pulpy part of the fruit and in the skin.

Rutin for Blood Vessels

While rutin is found in citrus fruits, it's also abundant in buckwheat, a grain that is seldom used in the

United States but popular in other parts of the world. Cherries, apricots, grapes, and plums are also good sources of rutin.

Rutin is useful for treating poor circulation and high blood pressure, and can help with hemorrhoids and varicose veins by preventing recurrent bleeding due to weakened blood vessels. A condition known as chronic venous insufficiency (CVI) can be treated with a derivative of rutin known as hydroxyethylrutoside (HR). CVI sometimes occurs after excessive clotting and inflammation of the leg veins, a disease known as deep vein thrombosis.

Chronic Venous Insufficiency (CVI)
When valves in leg veins cannot counteract gravity and keep blood in the legs, blood movement out of the veins slows, resulting in leg swelling.

Recent research has shown that rutin has other health benefits. For example, scientists in Argentina tested the anti-inflammatory properties of rutin, quercetin, and hesperidin. They found that all three were effective, but rutin was most active in fighting chronic inflammation. Furthermore, a clinical trial in Spain found that rutin protected the stomach cells from damaging free radicals, and in India, rutin was found to protect brain cells from the type of free-radical damage that occurs during a stroke.

Ellagic Acid Gets an A

For many people, the most memorable part of eating red raspberries is the tiny seeds that get stuck in the teeth. But scientists are proving that the delicate red raspberry is literally packed with health benefits, thanks to its high ellagic acid content. A number of clinical trials at Hollings Cancer Center at the University of South Carolina, for example, have shown that red raspberries, one of the richest food sources of ellagic acid, inhibit the growth of cancer cells. Moreover, these researchers found that ellagic acid arrested the growth of cancer in people with a genetic predisposition for the disease.

Daily consumption of one cup of red raspberries slows the growth of abnormal colon cells in humans, and has been shown to prevent the development of—and in some instances destroy—cells infected with human papilloma virus (HPV), believed to be the primary cause of cervical cancer. But that's only the beginning of the story.

Ellagic acid protects cells' p53 gene, which is in charge of supervising normal cell division and defending cells from invasion by cancerous cells. It also inhibits a stage in the growth cycle of cells known as mitosis that can cause cancer to spread. In tests, ellagic acid performed these protective duties with cervical, prostate, breast, esophageal, skin, colon, pancreas, and leukemia cells.

Mitosis
Ordinary division of a cell to form two new cells, identical to the parent cell.

At Ohio State University, black raspberries, which have exceptionally high levels of ellagic acid, have been tested against cancer with similar results. A variety of black raspberries known as "Blackcaps" protected lab animals against esophageal, colon, and oral cancer. Researchers noted that as the fruit moves through the digestive tract, its compounds are absorbed by various organs. The berries' high concentrations of antioxidants not only reduce cancer risk but combat inflammation, as well. As impressive at it is, ellagic acid is very likely not the only reason these berries are good at what they do, say the Ohio State scientists, who are hoping to identify other helpful substances soon. (Note that black raspberries are not the same as blackberries, but research is showing that all berries contain healthy compounds in varying amounts.)

Ellagic acid provides other services. It acts as a scavenger, binding with cancer-causing chemicals and inactivating them. It also inhibits other chemicals from causing mutations in bacteria and protects DNA against damaging carcinogens.

Raspberries (red and especially black), boysenberries, strawberries, walnuts, pecans, and cranberries are among the top sources of ellagic acid. Not surprisingly, epidemiological studies show that people who consume fruits high in ellagic acid have lower rates of cancer and heart disease.

Most flavonoids are not yet household names. But don't let their lack of fame prevent you from taking advantage of the many benefits they have to offer. Try a cup of green or black tea instead of coffee, have an orange for a midafternoon snack, or spread some raspberry jam on a piece of whole wheat bread. You're not only treating your taste buds—you're providing your whole body with the nutrients it needs to stay healthy.

CHAPTER 6

THE DIET DILEMMA: FOOD VERSUS SUPPLEMENTS

A few years ago, the National Cancer Institute (NCI) began the "Five a Day" campaign, designed to encourage people to eat at least five servings of fruits and vegetables a day. Although medical experts believe as much as one-third of all cancer cases could be related to poor diet, most Americans have not heeded these recommendations.

Studies have found that somewhere between 9 and 32 percent of Americans eat three to five daily servings of fruits and vegetables. So even in the most optimistic scenario, two out of every three Americans do not eat enough fruits and vegetables. In the worst-case scenario, nine out of ten people do not eat enough fruits and vegetables. Furthermore, most people who do eat fruits and vegetables eat from a very narrow group of these foods, so they lack the diverse selection of carotenoids and flavonoids that contribute to good health.

How about you? Did you eat five to nine servings of fruits and vegetables yesterday? Did you make certain to get something from the variously colored groups—the reds, purples, dark greens, yellows, and oranges? If not, you may have loaded up on one particular substance, such as lycopene from various forms of tomatoes, or lutein from lots of greens, and gotten very little of the others.

There's no doubt that lycopene and lutein are beneficial, but as we've seen, so are the other carotenoids and flavonoids. Unfortunately, most of us are so busy that there's little time for cooking. So we tend to settle for less-than-nutritious packaged meals and fast foods that contain few of the nutrients we've been discussing in this book. The good news is that it doesn't take much to turn poor eating habits around, if you keep a few things in mind.

The first point to remember is that when experts talk about servings, they are actually referring to very small amounts. In fact, a serving of fruits and vegetables consists of one medium-sized fruit, such as an apple or banana; half a cup of fruits or vegetables; or three-quarters of a cup of 100 percent fruit or vegetable juice.

If you have measuring cups handy, take a good look at the half cup—it's not much at all! Neither is one medium piece of fruit or three-quarters of a cup of juice. So enjoying a glass of orange juice for breakfast, an apple for a midmorning snack, and a big salad with assorted veggies at lunch could add up to five servings. Continue that approach into the evening, with another salad and a couple of half cups of vegetables with supper, plus some fruit for dessert, and you've hit the seven-serving mark—if not exceeded it—without any radical changes.

If you're thinking that the extra portions in the evening are a problem, here's the second thing to remember—vegetables are incredibly easy to cook. The best approach is to simply steam them in a saucepan with a very small amount of water. If possible, it's probably best to avoid microwaving. At least one study has shown that microwave cooking destroys much of the nutritional value of broccoli, for example, so until more is known, stick with steaming.

The Raw and the Cooked

Of course, some vegetables don't require cooking at all. Lettuce is a well-known example, and carrots, tomatoes, cauliflower, radishes, scallions, peppers, and many others can be eaten raw. Of course, some of these foods—like carrots and tomatoes—provide more healthful substances when they are cooked with a little oil, but even raw they still impart some of their goodness.

Soup is another way to increase your vegetable intake. Spend an hour or so one day cooking a favorite soup recipe, then freeze small portions for an easy meal during the next few weeks. The great thing about vegetable soup is that you can start with a basic recipe, and then experiment by adding your favorite vegetables in place of those you're not fond of. Another advantage is that you can add healthy but seldom-eaten vegetables—like kale or parsnips—to vegetable soup as a way of sampling them in flavorful surroundings. Soup also retains the nutrients often discarded in cooking water, so you're getting even more nutrition than in a serving of steamed veggies.

Even foods we consider not so healthy can be improved with the addition of vegetables. Pizza is a good example, because extra vegetables can be added to the standard toppings. When eating out, order a cup of vegetable or minestrone soup as a first course, instead of salad. You'll get the benefits of a variety of carotenoids and flavonoids that most restaurant salads can't provide. Or skip the bread, rice, or french fries and ask for side orders of veggies. Most restaurants now offer these as low-cost alternatives, even if they're not on the menu.

And don't forget that many commonly used herbs and spices are full of flavonoids, too. A recent study from the Fred Hutchinson Cancer Research Center in Seattle, published in the *American Journal of Clinical Nutrition,* noted that spices have the same cancer-protective benefits as vegetables, and can be antibacterial, anti-inflammatory, and antiviral, as well. Rosemary, basil, cilantro, oregano, turmeric, cinnamon, and many other spices are excellent sources of flavonoids, and terrific additions to meals, too.

Go Organic

To really ramp up your nutrient intake, choose organic fruits and vegetables whenever possible.

Although the subject is still controversial, several studies have shown that organic produce contains higher levels of vitamin C and polyphenols—the umbrella term for the carotenoid/flavonoid family—as well as other nutrient groups.

Organic

A specific method of growing and processing foods. Organic food is grown, packaged, and stored without the use of synthetic fertilizers, pesticides, herbicides, or irradiation.

It makes sense that organic produce would contain more carotenoids and flavonoids. In Chapter 1, we learned that these substances' original purpose was to protect plants from pests and other threats to their health. Organically grown fruits and vegetables are not treated with pesticides, so they must manufacture their own form of protection with greater amounts of carotenoids, flavonoids, and related compounds. Logically, that translates into higher levels of these substances for those of us who eat them.

Finally, don't forget that fruits and vegetables aren't the only sources of these nutrients. Beans, legumes, grains, tea, spices, and nuts contain flavonoids, too, and some seafood—particularly salmon and shrimp—provide the carotenoid astaxanthin. Levels of the nutrient can vary, though, depending on the fish's "lifestyle." For example, farm-raised Atlantic salmon can contain anywhere from 4 to 10 mg of astaxanthin per kilogram, but the levels in wild Pacific salmon are much higher, according to a recent FDA survey, which found that coho salmon contained approximately 14 mg/kg and sockeye salmon provided a whopping 40 mg/kg.

Here's how that translates into servings: A typical portion of four ounces of Atlantic salmon would supply about 1 mg of astaxanthin, while the same-sized serving of wild sockeye salmon would provide about four times as much. (Vegetarians will be happy to hear that in supplement form, algae, not fish, is the typical source of astaxanthin.)

Don't feel guilty if your diet doesn't measure up to recommended standards of fruit and vegetable intake. These once staple foods are not all that easy to come by in our culture. Then there's the fact that some people just don't like the taste of particular vegetables. For example, scientists have determined that some people's genes make them "supertasters," so they are more sensitive to bitter tastes found in some vegetables such as broccoli.

Of course, even the die-hard vegetable lover will find there are some days when cooking is out of the question, and even getting a decent meal seems next to impossible. That's when supplements are a great alternative.

Can supplements make up for a diet lacking in fruits and vegetables? Scientists really can't answer this question yet, but research is ongoing. Right now, experts believe that various types of carotenoid and flavonoid supplements can partly make up for a poor diet. And studies have found that supplements can reduce the risk of disease in people who do not usually eat fruits and vegetables.

Generally speaking, it's unrealistic to expect supplements to provide all of the beneficial nutrients found in whole foods. But when it comes to how well our bodies absorb those nutrients, carotenoids are a special case. Carotenoids in supplements are actually better absorbed than those in foods. The reason? Carotenoids in foods are locked in a fibrous matrix that is difficult for the body to break down. Cooking foods helps increase carotenoid absorption, but be careful not to overcook or you'll destroy these nutrients. (Overcooked vegetables tend to be limp and soggy. With most vegetables, it's better to undercook a bit, so they still have a little "crunch" to them.) The carotenoids in dietary supplements do not contain a fibrous matrix, though, so they are very well absorbed, as several scientific studies have demonstrated.

All in all, eating a diet filled with fruits, vegetables, whole grains, and nuts should be everyone's goal. Given the fact that most of us are simply not able to achieve this goal, however, it makes sense to use supplements as backup. In the next chapter, we'll look at smart ways to do just that.

CHAPTER 7

BUYING AND USING CAROTENOIDS AND FLAVONOIDS

There are many different types of carotenoid and flavonoid supplements on the market, enough to make shopping confusing. As with all purchases, it's important to be an educated consumer. In this chapter, you'll find suggestions for buying carotenoids and flavonoids and deciding which products might be the best ones for you.

Keep in mind as you read the following dosage instructions that they are intended for the average healthy adult. If you are taking medication of any kind, or are pregnant or nursing, you should not take supplements without consulting your medical practitioner.

Carotenoids

Individual carotenoids can enhance relatively specific aspects of health. As we have seen, beta-carotene, for example, can enhance the immune system, increase resistance to sunburn, and increase lung capacity. It is also a safe, nontoxic source of vitamin A, because our bodies can convert it to that nutrient if necessary. Lutein is essential for vision and likely reduces the risk of macular degeneration. Lycopene can reduce the risk of prostate cancer. If you are at risk of developing one of these diseases, it may be prudent to take an individual carotenoid supplement.

There are several major types of carotenoid supplements available, as well as "mixed carotenoids," which contain various natural carotenoids, usually derived from algae or palm oil. Individually, beta-carotene is found in both natural and synthetic forms. Typically, the source of natural beta-carotene is a type of algae, such as *Dunaliella salina*, and this is usually identified on the label. Natural lutein (from marigold petals) and lycopene (from tomatoes) are also found in dietary supplements. Since beta-carotene is one of the best-known and most readily available individual carotenoids, let's look at it first.

Natural versus Synthetic Beta-Carotene

To understand how natural beta-carotene is different from its synthetic cousin, we need to get a bit technical. The differences are significant, though, and worth knowing.

Natural beta-carotene consists of two molecules called isomers. One is 9-cis beta-carotene; the other is all-trans beta-carotene. Both contain the same molecules, but those molecules are arranged differently, just as two bunches of grapes contain basically the same ingredients but not arranged in the same way.

According to a noted beta-carotene researcher at Israel's National Institute of Oceanography, the natural 9-cis form of beta-carotene is the primary antioxidant (and protective) part of the molecule. The same scientist found that the 9-cis form of beta-carotene formed a substantial portion of total beta-carotene in lettuce, parsley, sweet potatoes, and many other foods. In contrast, synthetic beta-carotene consists of only the all-trans isomer, a very weak antioxidant. Synthetic beta-carotene contains no 9-cis beta-carotene. Many studies have shown the natural 9-cis form of beta-carotene to be more effective than the synthetic form.

This brings us back to the studies with lung cancer and beta-carotene supplements in Chapter 2. Because they used synthetic beta-carotene, what those studies really found was that the synthetic form might increase the risk of lung cancer under some circumstances, such as in people who are heavy smokers and drinkers. As for natural beta-carotene, those studies revealed nothing.

To find natural beta-carotene, look closely at the labels of beta-carotene, mixed carotenoid, anti-oxidant formulas, or multivitamin supplements. The richest supplemental sources generally come from algae, which might be identified on the back of the label as *Dunaliella salina, D. salina,* or some related type of algae. These types of algae are specially grown to be rich in beta-carotene. They naturally comprise 40 to 50 percent of the desirable 9-cis beta-carotene and around 50 percent of natural all-trans beta-carotene, with small amounts of other carotenoids, including alpha-carotene, lutein, zeaxanthin, and cryptoxanthin.

As for dosage, the average recommendation is for 15 mg (25,000 IU) daily for most adults. Smaller amounts can be beneficial, too, though. In a study in the *American Journal of Clinical Nutrition,* researchers calculated that for women, 5.37 mg (8,950 IU) of beta-carotene could protect against LDL oxidation. In some cases, however, more is better. Anyone with cataracts or macular degeneration should take two 15 mg doses of mixed carotenoids a day.

A Note about Dosage

For many years, beta-carotene was seen only as a precursor to vitamin A. Vitamin A measurements are always given in international units (IU), and so beta-carotene was measured in terms of its equivalence to vitamin A (and sometimes to "retinol units"). Now beta-carotene is being recognized for its non-vitamin-A antioxidant properties. As a consequence, more companies are listing amounts in milligrams (mg).

If necessary, there is an easy way to convert from mg to IU and vice versa. Basically, 5,000 IU is the same as 3 mg of beta-carotene.

For more complex conversions, it helps to have

a calculator handy. If you have a product listing beta-carotene in IU and want to know the amount of mg, multiply the number of IU by 0.0006. For example, one carrot contains about 20,000 IU, or 12 mg, of beta-carotene. Conversely, if your product lists beta-carotene in mg and you want the number of IU, divide the number of mg by 0.0006. For example, 12 mg divided by 0.0006 equals 20,000 IU. The IU/mg issues apply only to beta-carotene. Lutein and lycopene should always appear as mg because the body cannot convert them to vitamin A.

Lutein Supplements

Like beta-carotene supplements, lutein is available over the counter in a number of different products. Generally, lutein is extracted from marigold flower petals, a naturally rich source of the nutrient. There is some zeaxanthin in these supplements, and the body will make a little more from the lutein.

There are two types of lutein supplements on the market, and both are derived from marigold petals, a rich natural source of lutein. Some of these lutein products are, in fact, marketed as "marigold extracts." One type of lutein consists of "free lutein"— pure, cystalline lutein. The other type is "lutein ester." The ester form of lutein occurs naturally in many common foods.

Ester
A type of compound that adds chemical stability and a longer shelf life to natural products.

Humans absorb both free lutein and lutein ester very well. There is some preliminary evidence, based on research conducted at the University of Illinois, that the lutein ester may be assimilated and retained a little better than free lutein.

For general health maintenance, 4–6 mg of lutein daily should be sufficient. If you are trying to enhance protection against macular degeneration, 30–40 mg daily might be helpful. Because lutein is a fat-soluble nutrient, you can enhance its absorption by eating it with a tiny bit of an oily or fatty food, or just a regular meal.

A Look at Lycopene Supplements

A large ripe tomato contains about 4 mg of lycopene. This amount, on a daily basis, seems to be very good for health. As studies have shown, greater amounts of tomato foods reduce the risk of disease. There does not seem to be any harm from taking higher doses of lycopene, which is extracted from tomatoes.

According to a recent report on lycopene in *Alternative Medicine Review*, therapeutic dosages range from 6 to 60 mg daily, depending on the purpose. For example, 6 mg is recommended for reducing the risk of prostate cancer. Meanwhile, 6.5 mg appears to be sufficient to lower the risk of lung cancer in nonsmoking women, and 12 mg is suggested for nonsmoking men. To decrease the growth of prostate cancer, 30 mg daily is the recommended dose. The same amount prevents exercise-induced asthma, but it takes twice that amount—60 mg—to reduce LDL cholesterol levels.

Mixed Carotenoids Make Sense

Unless you have a specific health condition or a doctor's recommendation to focus on only one carotenoid, mixed carotenoids might be the best choice. Although it's a little confusing, research suggests that high doses of beta-carotene might lower or wash out lutein. There is also evidence that taking only beta-carotene increases the body's levels of at least some other carotenoids. In a two-year-long study, researchers found that beta-carotene supplements increased levels of alpha-carotene and lycopene, for reasons that aren't clear. We need to learn more about how high doses of one carotenoid may or may not affect others. But in the meantime, mixed-carotenoid supplements and a diet rich in fruits and vegetables may be the best way to go.

Carotenemia
Temporary yellowing of the skin due to excessive beta-carotene intake.

Another minor side effect, which can occur from eating a lot of carrots or taking very high doses of beta-carotene and other carotenoids, is carotenemia.

This is not in any way harmful. If the color bothers you, simply reduce your carotenoid intake, or stop taking carotenoid supplements for a month or two and then resume them at a lower dose.

To be on the safe side, heavy smokers (more than one pack daily) or regular drinkers (more than one glass of liquor daily) are probably better off taking beta-carotene or carotenoids as part of a broader supplement program, including vitamins C and E and the mineral selenium.

As for alpha-carotene and cryptoxanthin, they are best obtained either through diet or via a mixed-carotenoid supplement. Astaxanthin is the only carotenoid we have considered that is not usually included in mixed-carotenoid supplements. The recommended dosage of individual astaxanthin supplements is 1 mg, taken twice daily.

Getting the Good Stuff from Green Tea

The easiest way to get the health benefits of green tea's EGCG is by drinking green tea. If you're not excited about its slightly bitter flavor, check the shelves at your favorite store again. There are many new flavored versions of green tea available now that are much more palatable to American tastes. If caffeine is a problem, rest assured that decaffeinated green tea is just as beneficial as the regular kind.

Of course, green tea supplements are also an option, and they pack quite an antioxidant punch. A cup of green tea typically contains 8 to 12 percent of the important antioxidants—like EGCG—but the supplements may provide 50 to 90 of the same substances. If you can't be bothered brewing green tea all day, this is good news. Tea drinkers need a daily dose of three to four cups (or more!), while 100 mg of green tea extract can supply the same amount of antioxidants. In the end, it all boils down to whether or not you're a tea drinker.

Do Isoflavone Supplements Work?

Soy foods are an excellent source of isoflavones, but supplements also are available. Which is better? That's really up to you. But be aware that health experts stress the importance of eating soy foods, and fortunately, there are many to choose from, including tofu, soybeans, tempeh, soy milk, and protein powder that mixes well in smoothies.

These foods provide not only the most predominant soy isoflavones—genistein and daidzein—but other substances, as well. And since there is no guarantee that genistein or daidzein work alone in promoting good health, it only makes sense to go the food route. Soy foods are also a good source of protein and calcium, and organic soy products are widely available.

Of course, some people are less than thrilled with soy foods in any form. If that's the case, supplements may solve the problem. Capsules or tablets containing extracted soy isoflavones are widely available. Many products provide a mixture of isoflavones, including genistein and daidzein. Just as mixed-carotenoid supplements make sense because they promote interactions and synergy among the substances, mixed isoflavones are probably the best choice for most people. Sixty mg of isoflavones daily is typically recommended, but higher doses may be prescribed by a health practitioner. Large doses of soy supplements could affect thyroid function, however. Individuals with hypothyroidism should consult a physician before using supplemental soy.

Hypothyroidism *Deficiency of thyroid hormone production in the thyroid gland.*

Ipriflavone, a synthetic soy isoflavone, is another option. In clinical trials, ipriflavone (7-isopropoxy-isoflavone) has been shown to have similar health benefits to soy. For example, in an Italian study of more than 450 middle-aged women who had experienced loss of bone density, the participants who took 600 mg of ipriflavone along with a daily calcium supplement maintained bone density, while those who took calcium alone experienced continued and significant bone loss.

The bone-strengthening abilities of ipriflavone are enhanced by taking both vitamin D and calcium with it. Women taking the prescription drug calcitonin to prevent bone loss may also boost its effectiveness by taking ipriflavone. One caution: Ipriflavone may interact negatively with theophylline, a prescription medication used to open up the airways in people with asthma and other breathing difficulties. If you are taking this type of medication, check with your physician before adding ipriflavone.

Supplementing with Quercetin

For general health, experts recommend 125–400 mg of quercetin daily. Specific ailments may require more. For example, asthma sufferers may need 250–500 mg three times daily for relief. For heartburn, a combination of 500 mg of quercetin and bromelain may be most effective, taken three times per day.

Some health authorities say quercetin should be combined with vitamin C, either as part of a formula or with individual supplements. Others recommend taking quercetin about twenty minutes before meals for best results.

There have been reports of interactions between quercetin and some blood pressure medications. If you are being treated with prescription medication for high blood pressure, check with your physician before taking quercetin.

Promoting Health with Proanthocyanidins

A healthy adult should take 50–100 mg of Pycnogenol or grape seed extract daily. For therapeutic purposes, try 100 mg three times daily. Dozens
of studies have shown Pycnogenol and grape seed
extract to be safe, and there are no known side
effects with recommended dosages.

Some health experts recommend varying dosages to combat specific conditions. For example,
to treat blood platelet stickiness that leads to
clogged arteries, smoke exposure, swelling or pain
caused by disorders of the veins, or diabetic
retinopathy, take 0.5–1 mg for every pound of body
weight. As symptoms subside, reduce the dosage
to 50–100 mg daily.

Pycnogenol is a patented formulation, which
means the contents of the supplements are pre-
dictable and uniform. But grape seed extract supplements may vary in strength, so look for supplements that are standardized to contain 92 to 95
percent procyanidolic oligomers (PCOs)—another
term for proanthocyanidins.

Superior Health with Citrus Flavonoid Supplements

Citrus flavonoids are an excellent choice for anyone
who gets stomach pain from the acid in citrus fruits.
A typically recommended dose for a healthy person
is 1,000 mg (or 1 g), one to three times daily.

A good flavonoid product should contain some
pure rutin or pure hesperidin, or both. Mixed citrus-
bioflavonoid complexes without these compounds
tend to be less expensive, but they may not be as
effective, either.

Flavonoids should be taken with vitamin C whenever possible, to maximize the benefits of both. If
you find that the formulas combining flavonoids
and vitamin C are too steep for your budget, pur-
chase individual vitamin C and flavonoid supplements as an alternative.

A few cautions: Several recent studies have shown that taking antioxidants, including flavonoids, while undergoing chemotherapy and/or radiation treatments may decrease the effectiveness of these therapies. Hold off on taking antioxidants until the breaks between chemotherapy or radiation.

Also, citrus flavonoid combinations with naringen (a flavonoid found in grapefruit and grapefruit juice) should not be taken when using calcium channel blockers to lower blood pressure. Possible interaction between naringen and the drug could cause blood pressure to fall dangerously low. Also, naringen should not be taken by anyone currently on an immunosuppressant drug. If you are concerned about possible drug interactions with citrus flavonoids, ask your physician or pharmacist for more information.

Renewing Health with Rutin

The recommended daily dose of individual rutin supplements ranges from 120 mg to more than 3,000 mg (3 g) daily, depending on the overall health of the individual and the condition being treated. For example, blood vessels can improve quickly and dramatically with rutin, sometimes in only a few weeks. Very high doses may cause diarrhea, so it's best to start on the low end of the dosage recommendations and work up to higher amounts gradually.

Because rutin tends to reduce blood platelet stickiness, individuals taking blood-thinning medication should consult a physician before starting rutin supplements.

Evading Cancer with Ellagic Acid

Studies have shown that eating one cup of red raspberries daily can slow the growth of abnormal colon cells in humans, and prevent and sometimes destroy development of cells infected with human papilloma virus, the cause of cervical cancer.

Apples, blackberries, pomegranates, and straw-berries are other good sources of ellagic acid. But be careful about how these foods are processed. Ellagic acid can withstand freezing and freeze-dry-ing, but heat ruins it. So pasteurized apple juice or applesauce will not have nearly as much ellagic acid as a fresh apple, for example.

As for supplements, a daily dose of 40 mg of ellagitannin, which converts to ellagic acid in the body, is recommended. Ellagic acid has a synergis-tic relationship with quercetin, according to one recent study, so it would be wise to take the two together. Red raspberry leaves also contain ellagic acid, and are available as supplements. Pregnant women should avoid supplements made with rasp-berry leaf, though, because it may induce labor. Other than that, ellagic acid seems to be quite safe.

Carotenoids and flavonoids can and should be taken as part of a comprehensive supplement and health maintenance regimen. Nutrients work best as a team, and the effect is always greater when there are more members on the team.

Most people should start with a high-potency multivitamin supplement and a separate multimin-eral supplement. (Minerals take up more space than vitamins, so trying to get everything in one supplement often shortchanges something.) To this regimen, consider adding mixed carotenoids, flavonoids with vitamin C (aim for at least 1,000 mg daily), and enough vitamin E to get 400–800 IU daily. Remember, supplements aren't intended to replace a nutritious diet, but they can provide extra insurance and protection on those days when eat-ing healthy is difficult at best.

CHAPTER 8

ON THE
HORIZON

Getting plenty of fruits and vegetables is a great way to protect your health and nourish your body. As we've seen, so many carotenoids and flavonoids work in conjunction with one another that it seems best to get them as nature intended—from food. Not only are they naturally combined with some of their strongest partners this way, but there's also a very real possibility that other health-promoting substances exist in foods that haven't been identified yet.

In fact, we've only scratched the surface of the health benefits of carotenoids and flavonoids. There are many other substances currently being studied in laboratories all over the world, with very promising results. Pomegranates, for example, have been found to contain more flavonoid antioxidants than green tea or red wine. And speaking of red wine, two of its flavonoid components—resveratrol and saponins—are also turning out to be powerful allies when it comes to maintaining good health. Surely there will be other new discoveries like these in the years ahead.

In the meantime, look for ways to get more of the goodness of fruits and vegetables into your daily diet. Although five servings a day is the recommended minimum, remember that serving sizes are actually quite small and steaming is a quick, efficient way to add a few vegetable side dishes to regular meals. On days when time is tight, enjoying between-meal snacks of raw fruits and veggies, plus a glass of juice or two, is an easy way to ramp up intake.

There is substantial evidence that carotenoids and flavonoids can protect us from some of the nation's top killers, including heart disease, cancer, high blood pressure, diabetes, obesity, and more. Once you move away from the confines of the traditional American meat-and-potatoes menu, and start exploring the amazing varieties of fruits and vegetables on the market these days, you'll discover that good taste equals good health. And there is no better time to start than right now!

CONCLUSION

Carotenoids and flavonoids have been essential ingredients in pre-human and human diets during millions of years of evolution. Many researchers believe that because of this co-evolution, we are dependent on these nutrients for optimal health. Both carotenoids and flavonoids enrich our lives with color, flavor, and texture, as well as important health benefits. Many foods contain these substances, so the best—and easiest—way to obtain a diverse selection is by eating a wide range of fruits, vegetables, beans, and other plant foods.

Although carotenoids and flavonoids are not technically recognized as essential nutrients, there is plenty of solid scientific evidence that they are highly beneficial to our health. Research has shown that they have a broad range of effects, including enhancing the immune system and reducing the risks of cancer and heart disease. Unfortunately, most Americans do not eat the recommended five to nine daily servings of fruits and vegetables.

Granted, it may take a little extra effort to increase your intake of fruits, vegetables, and other plant foods. But today's supermarkets and restaurants offer a tremendous range of produce all year long. Even some fast-food chains are getting into the act by adding salads to their menus. In short, there are few excuses for not getting plenty of carotenoids and flavonoids—especially now that you know how exceptionally good for us these nutrients can be.

SELECTED
REFERENCES

Agarwal, C, Singh, RP, Dhanalakshmi, S, et al. "Anti-angiogenic efficacy of grape seed extract in endothelial cells." *Oncol Rep* Mar 2004; 11(3): 681–685.

Battinelli, L, Mengoni, F, Lichtner, M, et al. "Effect of limonin and nomilin on HIV-1 replication on infected human mononuclear cells." *Planta Med* Oct 2003; 69(10):910–913.

Bennedsen, M, Wang, X, Willen, R, et al. "Treatment of H. pylori infected mice with antioxidant astaxanthin reduces gastric inflammation, bacterial load and modulates cytokine release by splenocytes." *Immunology Letters* Dec 1, 1999;70(3):185–189.

Borska, S, Gebarowska, E, Wysocka, T, et al. "Induction of apoptosis by EGCG in selected tumour cell lines in vitro." *Folia Histochem Cytobiol* 2003; 41(4):229–232.

Buchler, P, Reber, HA, Buchler, MW, et al. "Anti-angiogenic activity of genistein in pancreatic carcinoma cell is mediated by the inhibition of hypoxia-inducible factor-1 and the down-regulation of VEGF gene expression." *Cancer* Jan 2004; 1(1): 201–210.

Buttemeyer, R, Philipp, AW, Schlenzka, L, et al. "Epigallocatechin gallate can significantly decrease free oxygen radicals in the reperfusion injury in vivo." *Transplant Proc* Dec 2003; 35(8):3116–3120.

Casto, BC, Kresty, LA, Kraly, CL, et al. "Chemoprevention of oral cancer by black raspberries." *Anticancer Research* Nov–Dec 2002; 22(6C):4005–4015.

Chew, BP, Brown, CM, Park, JS, et al. "Dietary lutein inhibits mouse mammary tumor growth by regulating angiogenesis and apoptosis." _Anticancer Research_ Jul–Aug 2003; 23(4):3333–3339.

Chew, BP, Park, JS. "Carotenoid action on the immune response." _Journal of Nutrition_ Jan 2004; 134(1):257S–261S.

Chiba, H, Uehara, M, Wu, J, et al. "Hesperidin, a citrus flavonoid, inhibits bone loss and decreases serum and hepatic lipids in ovariectomized mice." _Journal of Nutrition_ Jun 2003; 133(6):1892–1897.

Chuwers, P, Barnhart, S, Blanc, P, et al. "The protective effect of beta-carotene and retinol on ventilatory function in an asbestos-exposed cohort." _American Journal of Respiratory and Critical Care Medicine_ Mar 1997;155(3):1066–1071.

Cramer, DW, Kuper, H, Harlow, BL, et al. "Carotenoids, antioxidants and ovarian cancer risk in pre- and postmenopausal women." _International Journal of Cancer_ Oct 1, 2001; 128–134.

Devaraj, S, Vega-Lopez, S, Kaul, N, et al. "Supplementation with a pine bark extract rich in polyphenols increases plasma antioxidant capacity and alters the plasma lipoprotein profile." _Lipids_ Oct 2002; 37(10):931–934.

Doostdar, H, Burke, MD, Mayer, RT. "Bioflavonoids: selective substrates and inhibitors for cytochrome P450 CYP1A and CYP1B1." _Toxicology_ Apr 3, 2000; 144(1–3):31–38.

Fuhrman, B, Elis, A, Aviram, M. "Hypocholesterolemic effect of lycopene and beta-carotene is related to suppression of cholesterol synthesis and augmentation of LDL receptor activity in macrophages." _Biochemical and Biophysical Research Communication_ Apr 28, 1997 ;233(3):658–662.

Heo, HJ, Kim, MJ, Lee, JM, et al. "Naringenin from CITRUS JUNOS has an inhibitory effect on acetylcholinesterase and a mitigating effect on amnesia." _Dement Geriatr Cogn Disord_ 2004; 17(3):151–157. Epub Jan 20, 2004.

Hu, G, Cassano, PA. "Antioxidant nutrients and pulmonary function: the Third National Health and

Nutrition Examination Survey (NHANES III)." *American Journal of Epidemiology* May 15, 2000; 151(10): 975–981.

Kritchevsky, SB, Tell, GS, Shimakawa, T, et al. "Provitamin A carotenoid intake and carotid artery plaques: the Atherosclerosis Risk in Communities Study." *American Journal of Clinical Nutrition* Sep 1998;68(3):726–733.

Kumar, NB, Cantor, A, Allen, K. "The specific role of isoflavones on estrogen metabolism in premenopausal women." *Cancer* Feb 15, 2002; 94(4): 1166–1174.

Lampe, JW. "Spicing up a vegetarian diet: chemopreventive effects of phytochemicals." *American Journal of Clinical Nutrition* Sep 2003; 78(3 Suppl):579S–583S.

Mannisto, S, Smith-Warner, SA, Spiegelman, D, et al. "Dietary carotenoids and risk of lung cancer in a pooled analysis of seven cohort studies." *Cancer Epidemiology Biomarkers Prevention* Jan 1, 2004; 13(1):40–48.

Montonen, J, Knekt, P, Jarvinen, R, et al. "Dietary antioxidant intake and risk of type 2 diabetes." *Diabetes Care* Feb 2004; 27(2):362–366.

Moreno, DA, Ilic, N, Poulev, A, et al. "Inhibitory effects of grape seed extract on lipases." *Nutrition* Oct 2003; 19(10):876–879.

Muhlhofer, A, Buhler-Ritter, B, Frank, J, et al. "Carotenoids are decreased in biopsies from colorectal adenomas." *Clinical Nutrition* Feb 2003; 22(1):65–70.

Narayanan, BA, Geoffroy, O, Willingham, MC, et al. "p53/p21(WAF1/CIP1) expression and its possible role in G1 arrest and apoptosis in ellagic acid treated cancer cells." *Cancer Letters* Mar 1, 1999; 136(2): 215–221.

Osganian, SK, Stampfer, MJ, Rimm, E, et al. "Dietary carotenoids and risk of coronary artery disease in women." *American Journal of Clinical Nutrition* Jun 2003; 77(6):1390–1399.

Ozasa, K, Nakao, M, Watanabe, Y, et al. "Serum phytoestrogens and prostate cancer risk in a nested case-control study among Japanese men." *Cancer Sci* Jan 2004; 95(1):65–71.

Peng, QL, Buz'Zard, AR, Lau, BH. "Pycnogenol protects neurons from amyloid-beta peptide-induced apoptosis." *Brain Research and Molecular Brain Research* Jul 15, 2002; 104(1):55–65.

Putter, M, Grotemeyer, KH, Wurthwein, G, et al. "Inhibition of smoking-induced platelet aggregation by aspirin and Pycnogenol." *Thrombosis Research* Aug 15, 1999; 95(4):155–161.

Rohdewald, P. "A review of the French maritime pine bark extract (Pycnogenol), a herbal medication with a diverse clinical pharmacology." *International Journal of Clinical Pharmacology Therapy* Apr 2002; 40(4):158–168.

Sarkar, FH, Li, Y. "Soy isoflavones and cancer prevention." *Cancer Invest* 2003; 21(5):744–757.

Sato, R, Helzlsouer, KJ, Alberg, AJ, et al. "Prospective study of carotenoids, tocopherols, and retinoid concentrations and the risk of breast cancer." *Cancer Epidemiology Biomarkers Prevention* May 2002; 11(5):451–457.

Schonlau, F, Rohdewald, P. "Pycnogenol for diabetic retinopathy: a review." *International Ophthalmology* 2001; 24(3):161–171.

Singh, RP, Tyagi, AK, Dhanalakshami, S, et al. "Grape seed extract inhibits human prostate tumor growth and angiogenesis and upregulates insulin-like growth factor binding protein-3." *International Journal of Cancer* Feb 20, 2004; 108(5):733–740.

Street, DA, Comstock, GW, Salkeld, RM, et al. "Serum antioxidants and myocardial infarction. Are low levels of carotenoids and alpha-tocopherol risk factors for myocardial infarction?" *Circulation* Sep 1994; 90(3):1154–1161.

Suzuki, K, Ito, Y, Nakamura, S, et al. "Relationship between serum carotenoids and hyperglycemia: a population-based cross-sectional study." *Journal of Epidemiology* Sep 2002; 12(5):357–366.

Toniolo, P, Van Kappel, AL, et al. "Serum carotenoids and breast cancer." *American Journal of Epidemiology* Jun 15, 2001; 153 (12):1142–1147.

Uchiyama, K, Naito, Y, Hasegawa, G, et al. "Astaxanthin protects beta-cells against glucose toxicity in diabetic db/db mice." *Redox Rep* 2002; 7(5): 290–293.

Wu, K, Erdman, JW Jr., Schwartz, SJ, et al. "Plasma and dietary carotenoids and the risk of prostate cancer: a nested case-control study." *Cancer Epidemiology Biomarkers Prevevention* Feb 1, 2004; 13(2):260–269.

Yamakoshi, J, Saito, M. Kataoka, S, Tokutake, S. "Procyanidin-rich extract from grape seeds prevents cataract formation in hereditary cataractous (ICR/f) rats." *J Agric Food Chem* Aug 14, 2002; 50(17):4983–4988.

Zheng, Y, Song, HJ, Kim, CH, et al. "Inhibitory effect of epigallocatechin 3-O-gallate on vascular smooth muscle cell hypertrophy induced by angiotensin II." *J Cardiovasc Pharmacol* Feb 2004; 43(2):200–208.

OTHER BOOKS
AND RESOURCES

Challem, J. *The Inflammation Syndrome: The Complete Nutritional Program to Prevent and Reverse Heart Disease, Arthritis, Diabetes, Allergies, and Asthma.* Hoboken, NJ: Wiley, 2003.

Duyff, RL. *American Dietetic Association Complete Food and Nutrition Guide, Second Edition.* Hoboken, NJ: Wiley, 2002.

Heber, David, M.D., Ph.D. *What Color Is Your Diet?* New York: HarperCollins, 2001.

GreatLife Magazine
Consumer magazine with articles on vitamins, minerals, herbs, and foods.
Available for free at many health and natural food stores.

Let's Live Magazine
Consumer magazine with emphasis on the health benefits of vitamins, minerals, and herbs.

Customer service:
1-800-676-4333
P.O. Box 74908
Los Angeles, CA 90004
Subscriptions: 12 issues per year, $19.95 in the U.S.; $31.95 outside the U.S.

Physical Magazine
Magazine oriented to body builders and other serious athletes.

Customer service:

1-800-676-4333

P.O. Box 74908

Los Angeles, CA 90004

Subscriptions: 12 issues per year, $19.95 in the U.S.; $31.95 outside the U.S.

The Nutrition Reporter™ newsletter

Monthly newsletter that summarizes recent medical research on vitamins, minerals, and herbs.

Customer service:

P.O. Box 30246

Tucson, AZ 85751-0246

e-mail: jack@thenutritionreporter.com

www.nutritionreporter.com

Subscriptions: 12 issues per year, $26 in the U.S.; $32 U.S. or $48 CNC for Canada; $38 for other countries.

USDA National Nutrient Database for Standard Reference

Search for nutrients in common foods at these websites:

www.nal.usda.gov/fnic/cgi-bin/nut_search.pl

www.nal.usda.gov/fnic/foodcomp/Data/SR16/
   wtrank/wt_rank.html

USDA's Beltsville Agricultural Research Center (BARC) has a page devoted to phytonutrients, with FAQs and links:

   www.barc.usda.gov/bhnrc/pl/pl_faq.html

INDEX

Printed in the USA
CPSIA information can be obtained
at www.ICGtesting.com
JSHW050801160824
68134JS00068B/78